CW00349181

The Modern British Data State, 1945–2000

This political history studies the phenomenal growth of the modern British state's interest in collecting, collating and deploying population data. It dates this biopolitical data turn in British politics to the arrival of the Labour government in 1964. It analyses government's increased desire to know the population, the impact this has had on British political culture and the institutions and systems introduced or modified to achieve this. It probes the political struggles around these initiatives to show that despite setbacks along the way and regardless of party, all British governments since the mid-1960s have accepted that data is the key to modern politics and have pursued it relentlessly.

Kevin Manton teaches History and Politics at SOAS and Birkbeck, London. He researches modern British history and is the author of *Population Registers and Privacy in Britain, 1936–1982* (2019).

Routledge Studies in Modern British History

The British Jesus, 1850–1970
Meredith Veldman

The Rise and Decline of England's Watchmaking Industry, 1550–1930
Alun C. Davies

The Football Pools and the British Working Class
A Political, Social and Cultural History
Keith Laybourn

Respectability, Bankruptcy and Bigamy in Late Nineteenth and Early Twentieth-Century Britain
John Benson

The Beveridge Report
Blueprint for the Welfare State
Derek Fraser

The Modern British Data State, 1945–2000
Kevin Manton

Labour's Ballistic Missile Defence Policy 1997–2010
A Strategic-Relational Analysis
Dr James Simpkin

For more information about this series, please visit: https://www.routledge.com/history/series/RSMBH

The Modern British Data State, 1945–2000

Kevin Manton

LONDON AND NEW YORK

First published 2023
by Routledge
4 Park Square, Milton Park, Abingdon, Oxon OX14 4RN

and by Routledge
605 Third Avenue, New York, NY 10158

Routledge is an imprint of the Taylor & Francis Group, an informa business

British Library Cataloguing-in-Publication Data
A catalogue record for this book is available from the British Library

ISBN: 978-1-032-17252-1 (hbk)
ISBN: 978-1-032-17254-5 (pbk)
ISBN: 978-1-003-25250-4 (ebk)

DOI: 10.4324/9781003252504

Typeset in Times New Roman
by codeMantra

Printed in the United Kingdom
by Henry Ling Limited

For Sam, Dan & Evelyn

Contents

Contents

Acknowledgements

This book was written using the resources of the UK National Archive at Kew, London. It would have been impossible without the good-natured professionalism of the staff there, and it is a pleasure to be able to thank them here. The staff and reviewers at Routledge have been hugely encouraging throughout this book's gestation and I am grateful for this. Similarly deserving of thanks, though for different reasons, are Oscar, Soni, Tim, Brett, Barry, Steve, Sam, Sabina, Gouldie and Gabriella. I would also like to take this opportunity to thank all my students for being such positive people and such great company even when the times we lived in seemed so overwhelmingly negative.

Introduction

In 2009, the Joseph Rowntree Trust issued a report, *The Database State*, which presented information about government databases and concluded that of the forty-six it surveyed only six could be labelled as unambiguously legal.[1] Some of the other forty represented developments initiated by the Blair government that was in power at that time, but, regardless of when any particular database was organised, this book argues that in order to grasp the nature of these government data streams it is necessary to appreciate two things. The first of these is when this desire for an increased depth and breadth of population data arrived at the centre of British politics because, though this has become a defining feature of all modern British governments, this has not always been the case.

It is noticeable that each of the systems analysed by the Rowntree report can be seen as a component of a wider net of data, the sum of which represented an attempt to develop a comprehensive knowledge of the British population. This means that these systems were developing something similar to the database called the National Register, which had been in use during the Second World War but which was abandoned in peacetime. This register was the data system behind wartime identity cards and held an up-to-date address for everyone in the country. Looking back from the publication of the Rowntree report, it seems astonishing that this wartime register was closed in 1952 and all the more so, because it was closed as no one in government circles at that point could see a need for it in peacetime. Clearly, as the Rowntree report highlighted, this attitude, and this perception of needs, has changed markedly, and this book aims to analyse and detail this change and to mark the point at which this change in government attitudes occurred. Commenting on this point, in a pair of academic articles published in 2005, a group of scholars dated the origin of these developments under Blair to the mid-1990s.[2] However, this book argues these systems originated in the 1960s with the accession of the 1964 Labour government. This came into office promising to initiate a "revolution" in government data systems and lived up to this promise by inaugurating a data turn in British government and politics.[3] It is, for sure, always possible to trace the roots of any contemporaneous developments back further into the past. Indeed some have sought

DOI: 10.4324/9781003252504-1

to argue that modern government data gathering is all of a piece with events going back to the Doomsday Book.[4]

However, contrary to this longer-term view, which tends to see processes as characterised by incremental changes, the view here is that British government attitudes towards population data went through a step change in the mid-1960s. Thus, this book shows that many of the key phrases that resonate through the Blair government's plans and innovations, such as the desires to modernise data systems, to integrate (join up) policy across government, to utilise government data holdings to the full or to deal with the public "in the round" as individuals, all entered the government's lexicon in the 1960s.[5] Moreover, the Blair governments' desire to use their data systems to crack down on social-security fraud and illegal immigration, along with their exclamations of despair at the limitations imposed on policy by their poor knowledge of the population, would also have been recognisable to their predecessors (though the focus on immigration and fraud was an addition to the rhetoric that gathered pace in the 1970s and 1980s).

The second thing that is emphasised in this book is that these connections to the past are not simply coincidences of language, rather the policies and language used under Blair are the same as those of earlier years because they emanate from the same intellectual-political milieu, characteristic of the drive for modernity, that underpinned both Wilson's and Blair's governments. This has been conceptualised by Michel Foucault as focusing on biopolitics. Foucault sees biopolitics as the sum total of attempts to exercise power not over individuals, but rather over people as a mass, in other words: the population.[6] Achieving this, as Bruce Curtis has noted, involves the government in gaining access to the abstraction of the population through increased access to the individuals who comprise it.[7] In other words, it involves the use of political data technologies such as censuses, surveys, registrations, licences and identity cards. The pursuit of the welfare, health and education policies that are at the centre of biopolitics is so crucially dependent on data that Foucault further argues that any modern state inevitably deploys data-gathering systems and indeed that a modern state cannot be conceived without such systems.[8] This view allows the development of an understanding of why all of these data technologies were either introduced, or reshaped, to achieve greater or different flows of information into government, or seriously discussed in government circles so often in the period, and indeed ever since.

However, this does not mean that all states are equally modern at the same time. Thus, it is important to note that having described the workings of the police in France and Germany as the embodiment of this new form of politics, Foucault added that in Britain the police represented a very different system and, elsewhere, he was at pains to contrast the system he described and analysed on mainland Europe to "a much looser Anglicized system."[9] It is this book's argument that, during the period it discusses, this looser system was subject to significant tightening through the processes of modernisation with their focus on biopolitical interventions driven by population data.

As such this is a work of history rather than theory, it looks at a particular example, Britain, in a particular period, 1945–2000, but such an approach need not be alienated from theory. Indeed, Nikolas Rose is clear that understanding how the system he conceptualises actually unfolded, "would be a matter for a certain type of empirical investigation," one that examined the complexities, contingencies, local variables and illogical twists of this process.[10] As a work of political history, this book aims to present this empirical process in order to detail how British governments attempted to move towards a modern, data-driven set of biopolitical policies. Thus, it further shows how, because these governments made these attempts within the relatively loose system they inherited, they were obliged to navigate a course around convoluted contingencies, such as the traditional patterns of data confidentiality and the institutional structures of the state, that were characteristically British.

Welfare is central to Foucault's understanding of modern power, and the way welfare exercises this function has been analysed by Nikolas Rose and Peter Miller. They argue that welfare seeks to realise its aims by creating an interconnected web of knowledge linking otherwise separate networks. This whole is designed to bring knowledge about the people to the centre of the state so that they can be observed and understood and thus acted upon. In order for these processes to function, the knowledge has to be transferred in a standardised format meaning that the people's lives need to be rendered into statistics. The three watchwords of the data system that Rose and Miller place at the heart of welfare are thus standardisation, centralisation and linkage, and this book determines that these were the three main data strategies pursued by British governments in this period.[11]

However, these broad strategic aims were not the same thing as an actual policy or system for data collection. The original plan developed by the Wilson government was to reintroduce a full-blown population register, or at least to introduce a common-numbering system for government data sets that, by linking together data sets that had always been held apart, would achieve more or less the same depth and breadth of vision over the population. However, this plan had to be withdrawn, which meant that the government was obliged to overhaul existing data-collection devices, the census or driving licences, for example, or to introduce new methods of gathering data, such as ID cards, in order to either develop a new overarching data base or draw together the information it held in its extant, discreet filing systems. But just as governments were obliged by austerity, popular pressure or the patterns of the traditional constitution and common law to adopt and adapt the systems they already had at their disposal, so they had to make additions to or reform, rather than radically overhaul, the institutions that actually operated these data systems. Thus, data reform developed through what Michael Moran has called a series of "shotgun weddings" with traditional structures.[12] This book thus examines the nature and the role of the government's data-gathering bodies such the Office for Population Censuses and Surveys (OPCS). It details how this body was set up at the end of the 1960s, within the Ministry of Health, to be in the vanguard of the new drive for data but, within seven years, had come to be seen across government as falling short of what was required and was eventually, in 1996, folded into the Central Statistical Office to create the Office for National Statistics. By examining institutional systems that collected and collated population data for the government in general, and the rise and fall of the OPCS in particular, this book charts the centralisation of data streams that was an intimate part of the data turn in British politics.

The policy fields examined here, for example, welfare, education and health, are intensely political fields, while data, along with the knowledge and power that flow from it, is central to modern state power. Thus, it is perhaps paradoxical that these technical devices or data-based systems engendered a de-politicisation of the policies they were applied to. This unfolded because data was usually seen as being a scientific and objective device that was thus pre-positioned as the only possible solution to a problem once modern thinking had framed such problems in biopolitical terms. But even when politicians and civil servants found that their preferred solution failed, or produced unforeseen outcomes, or as policy changed with a change in government, the solution they reached for was always the same: more data to give more knowledge to produce better policies. Thus, once the state applied a broadly biopolitical mind to issues concerning bettering, reforming or defending the population, it was inevitable that it would set off in pursuit of a chimera of a perfect knowledge of the population. Thus, through the period studied here, all governments found themselves dissatisfied with the data streams they inherited from their predecessors. They all increased the depth, breadth and pace of the processes of standardisation, centralisation and linkage of data systems along with the centralisation of data-gathering institutions. Nevertheless, and despite all this effort, they remained ever unable to resolve the issues they saw themselves facing. Governments' demands for data thus led to an accession to the demands of data.

As part of this approach, the modern British data state viewed any sign of hostility from the people as, on the one hand, being, at best, proof that the public did not understand what its true interests actually were. The people, in other words, needed modernising too. While, on the other hand, at worst, British bloody-minded opposition to being counted, catalogued or classified could be a sign that the people were up to something and could not be trusted. But regardless of which view prevailed, more data would be required to implement the required policy prescriptions and yet more would be required to protect the integrity of the systems that gathered, held and deployed this information. In this manner, the de-politicisation at the heart of the data state became a self-fulfilling prophecy.

Other works on this, or related subjects, take rather a different view of the growth of government data systems. Thus, the essays in the collections edited by Ilsen About, James Brown and Gayle Lonergan; Jane Caplan and

John Torpey; and Keith Breckenridge and Simon Szreter, tend to highlight what might be called the positive aspects of government population data by, for example, looking at how it allows people to participate in a state's politics and gives access to welfare, health and education systems, and in doing so, these essays cover a wide range of states.[13] As such, they tend to assume that government data is needed and therefore do not address the central questions faced by this book: how and why did this system develop in Britain after the mid-1960s? Shoshana Zuboff is clear that data collection threatens privacy and freedom, but her work is primarily about the United States and, more importantly, is an analysis of how private internet businesses began compiling and selling data about their customers, and an exposé of the extent to which they continue to do so. As such, it is not about government at all and in fact it tends to assume government is somewhat benign in these matters suggesting that it should stand up to the tech giants.[14] This book shows how the British government's interest in amassing data predates the widespread arrival of computers, never mind the advent of Google. Moreover, it shows how governments have used the technology both to develop their data capabilities and to mask what they were doing behind a veil of technology that many people did not understand.

In making these points what follows is divided into nine chapters. These are arranged in a broadly chronological framework with chapters covering different themes or aspects of the emergence of the data state. Chapter 1 provides an overview of the government's view of population data before the election of the Wilson government in 1964. It establishes the parameters of the norms and protocols that governed the use of government data. It also sets out the position of the forces, such as medical research and social science that sought access to this data showing how, whether they were operating inside the state or in civil society, they were largely unsuccessful until 1964. Thus, this chapter gives an overview of the political norms of Britishness, with respect to population data, that were responsible for the abolition of the wartime national register and which, as Foucault remarked, differentiated Britain from the rest of Europe, while the rest of the book shows how these norms were altered following 1964.

Chapters 2 and 3 cover the governments of the period 1964–79. The first of these looks at Wilson's Labour government. It shows how this government wanted to implement a data revolution in the form of a reintroduced population register. However, once this plan was thwarted, it was obliged to attempt to achieve its aims through the existing data institutions of the state. That meant it had to engage in what would become a protracted war of position against the prevalent norms of data confidentiality and the decentralised nature of the Government Statistical Service (GSS). Chapter 3 looks at examples of the methods used by government to bolster its data resources. The first of these saw government conducting surveys of the people to gather new information. Here the examples of the National Cohort Study and the General Household Survey are analysed. Second, the chapter explores the increased use of data already held by government departments. This involved government in circumventing the existing confidentiality codes that ring-fenced data allowing it to be used only for the purpose for which it had been obtained, and commensurately, in increasingly standardising the format of data to facilitate the processes of linking it into a wider and deeper matrix.

Chapter 4 introduces some of the key themes in data gathering that began with the arrival of the Conservatives under Margaret Thatcher in 1979. The first of these was the acceleration of the centralisation of the GSS already evidenced under previous governments since 1964. This chapter details the stages of this process leading up to the formation of the Office for National Statistics in 1996. The argument here shows that this centralisation of structures was propelled by a desire to centralise and link the data itself. This chapter also shows how it was under Thatcher's governments that data became focussed on monitoring and combatting both immigration and fraud and that these rhetorical devices gave the government's data drive some level of political legitimacy. It also illustrates how this foregrounding of immigration led to protests against the inclusion of questions on race/ethnicity and national origins in the 1981 census.

Chapter 5 looks at the people and their relationship to government data. This establishes how the British were not a nation of technophobes, but rather they were concerned about government amassing data on them in any format. In fact it was the government that allowed and encouraged any attention attracted by its data systems to fall on computers because here it could offer technical assurances to the people that all was well. Indeed, framing these serious political issues as technical problems, amenable to technical solutions, was a method of circumventing any political discussion with the public. This depoliticisation, which was inherent in the growth of the modern biopolitical data state, meant that the government could not justify these new systems to the public by promising democratic empowerment. Instead, the government overwhelmingly claimed that its systems would offer people increased convenience. Thus, a type of consumerist language came to dominate government discussions of these issues. Notwithstanding its desire to keep its data operations concealed from public view, the government was obliged to pass the 1984 Data Protection Act. This chapter analyses the nature of this act and how the government was able to brush aside the data protection registrar's criticisms of how credit reference agencies and NHS contract data sets misused people's personal information.

Chapters 6–8 look at government data systems. Chapter 6 surveys government's continuing attempts to link its datasets and bolster its data streams. First, it examines attempts to overhaul the electoral register in order to turn it into something very close to a full population register. The electoral register had always been compiled locally and the issue of centre–local relations also surfaced in the poll-tax registers that are also examined in this chapter. These both represented attempts to compile new data systems. However, this chapter also takes an overview of how these governments increasingly pursued a variety of schemes to link the disparate filing systems across Whitehall through common-numbering systems, and shows how this policy became increasingly the data gatherer's preferred metier. Chapter 7 looks at another attempt to introduce a new data stream in Britain, through the introduction of ID cards. This chapter indicates how this scheme had its roots in an attempt to reformat the driving licence, to adapt the photo-bearing design in use in Northern Ireland. The discussion here details how these plans snowballed to propose the introduction of a multifunction smart card for use by both public and private sector organisations. Chapter 8 looks at the Government Data Network, showing how this provided the architecture for an acceleration of the deepening and widening of links between government datasets. This chapter also discusses the Information Society Initiative (ISI). This was a plan developed under the government of John Major to promote the spread of information technology both in society and the state and is indicative of government's role in increasing the use of computers and IT across society. As part of the ISI, the government planned to deploy a computerised system, called *government.direct*, which would marshal a lot of its relations with the public into one unified system. This was presented to the public as offering consumerist-style benefits that would ease their transactions with state bodies, but within the government, it was acknowledged that this system would permit the recasting of all government population data into one matrix.

Chapter 9 builds on the findings of Chapter 5 and so returns to examining the attitudes of the British people. It begins by showing how the government promoted the growth and spread of IT as a safe means of gathering, storing and using data and then determines that the reality of government data security was very different. This chapter then reports on surveys of the British people's attitudes towards data gathering, in general, and computerisation, in particular, showing the levels of fear and hostility that were prevalent at the time. The chapter concludes by showing how the 1998 Data Protection Act was passed and analyses and details the government's unwillingness to sign up to the terms of the European directive that was the genesis of this legislation.

The conclusion draws out six main themes explored throughout the book to show how these were all continued under the Blair government that came into office in May 1997. These themes are: the nature of modernisation in the British state and the integral role of population data in these processes; the centralisation engendered by this pursuit of data; the reliance on technical solutions to bridge the perceived gaps in the government's knowledge of the people; the de-politicisation inherent in this data-driven approach and how this was given a huge fillip by the advent of IT solutions; the desire within government to protect its data and its data-gathering systems from what it labelled as fraud and, allied to this, its desire to enhance and use its data systems in its struggle to control immigration.

This book is based on research conducted in the UK National Archives using government papers. Such documentation for the Blair governments was not available for consultation at the time of writing, but this concluding section uses publications from the Blair governments' early years to show how these tendencies continued under these governments. Indeed, the argument is made that, given the Blair governments' deployment of the same data-driven, biopolitical approach to British politics, its adoption of the same methods and indeed the same language, as its predecessors, was virtually inevitable.

Notes

1 Ross Anderson, Ian Brown, Terri Dowty, Philip Inglesant, William Heath and Angela Sasse, *Database State: A Report Commissioned by the Joseph Rowntree Reform Trust Ltd* (York: The Joseph Rowntree Reform Trust Ltd, 2009), 2.
2 Perri, 6, Charles Raab and Christine Bellamy, (2005) "Joined-Up Government and Privacy in the United Kingdom: Managing Tensions between Data Protection and Social Policy. Part 1," *Public Administration* 83 (2005): 111–133; and "Joined-up Government and Privacy in the United Kingdom: Managing Tensions between Data Protection and Social Policy. Part 2," *Public Administration* 83 (2005): 393–415. See 'Part 1': 114.
3 TNA, CAB 139/742, *Draft Speech for the Royal Society of Statisticians Banquet on the Occasion of the 37th Session of the International Statistical Institute*, Sept. 10, 1969, 11.
4 See, for example, Edward Higgs, *The Information State in England: The Central Collection of Information on Citizens since 1500* (London: Red Globe, 2003); and, Rhodri Jeffreys-Jones, *We Know All about You: The Story of Surveillance in Britain and America* (Oxford: OUP, 2017).
5 TNA, RG 19/863, *Paper on Setting up and Operation of the Longitudinal Study*, 28 Feb. 1978, 4.
6 Michel Foucault, *Society Must Be Defended* (London: Penguin, 2004), 246.
7 Bruce Curtis, *The Politics of Population: State Formation, Statistics, and the Census of Canada, 1840–1875* (London: University of Toronto Press, 2001), 24.
8 Michel Foucault, *Security, Territory, Population* (New York: Picador, 2004), 277.
9 Michel Foucault, *The Birth of the Clinic* (London: Routledge, 1989), 48.
10 Nikolas Rose, *Powers of Freedom: Reframing Political Thought* (Cambridge: Cambridge University Press, 1999), 19 and 210.
11 Nikolas Rose and Peter Miller, "Political Power Beyond the State: Problematics of Government," *The British Journal of Sociology* 43 (1992): 173–203.
12 Michael Moran, *The British Regulatory State: High Modernism and Hyper-Innovation* (Oxford: OUP, 2003), 179.
13 Ilsen About, James Brown and Gayle Lonergan, eds., *Identification and Registration Practices in Transnational Perspective: People, Papers and Practices* (Basingstoke: Palgrave, 2013).
Jane Caplan and John Torpey, eds., *Documenting Individual Identity: The Development of State Practices in the Modern World* (Princeton: Princeton University Press, 2001).
Keith Breckenridge, and Simon Szreter, eds. *Registration and Recognition: Documenting the Person in World History* (Oxford: Oxford University Press, 2012).
14 Shoshana Zuboff, *The Age of Surveillance Capitalism: The Fight for a Human Future at the New Frontier of Power* (London: Profile, 2019).

1 The British Government's Approach to Population Data c1945–64

In 2003, David Blunkett, and the Labour government for which he was home secretary, attempted to introduce identity cards in Britain, and as part of this process, the government published the findings of a consultation exercise it had held. These stated that a fact, "highlighted by several people," who were in favour of ID cards, was that there had been "a card scheme in the war, which was viewed as successful."[1] Similarly, though from the other end of the party-political spectrum, in 1995 when Michael Howard attempted to introduce ID cards, he too was keen to show how "there have already been identity cards schemes in the United Kingdom" with his green paper then giving a page and a half over to a potted history of the schemes used in the two world wars.[2] This unity of purpose across the party divide to introduce ID cards tells the observer a lot about how amassing data on the population, and identifying the individuals who comprise it, has come to be an integral part of what the British government does. However, for now, it is important to highlight an issue that both these home secretaries implied but glossed over in their presentations.

This is the fact that what they each sought was the reintroduction of a scheme that an earlier holder of their office had seen fit to abolish. This earlier scheme was the National Register, the database that underpinned the national wartime systems of ID cards, conscription and food rationing. The purpose of this chapter is to briefly explore the reasons behind this abolition and to demonstrate how this was an integral part of the traditional British attitudes to population data that dominated the government until the mid-1960s. It is argued here that only by concentrating on these attitudes is it possible to understand the nature of the dramatic volte-face that occurred in the mid-1960s, when a new Labour government began to consider the reintroduction of the data system that had underpinned wartime ID cards, a process that would ultimately lead to Blunkett and Howard's positions. This chapter thus provides an overview of how the 1964 Labour government's thinking on population data was formed in the crucible of new ideas in the period up to its election when it would bring population data to the centre of British politics.

To do this, this chapter is divided into three sections. The first section examines the abolition of the wartime National Register to show how this

DOI: 10.4324/9781003252504-2

abolition was an integral part of a traditional British attitude that has been recognised by Michel Foucault as being at odds with practices in mainland Europe. The second and third sections look at social science and medical research respectively illustrating how, in the post-war period, governments kept them at arm's length and how this stymied both sides of this relationship, preventing the emergence of a fully modern biopolitical approach to politics and data, a logjam that was only broken by the arrival of the Wilson government in October 1964.

1

The ID cards that were abolished in 1952 were the physical representation of a data system called the National Register (the generic term is a population register). This register was an up-to-date list of everyone in the country designed to open the people up to the demands of conscription by the military or the wartime economy. To offer the government this service, the register needed to hold up-to-date addresses, and this was ensured through its being connected to the food-rationing system.[3] However, though the National Register was introduced as a wartime measure, both its pre-war creator and its main post-war defender, the registrar generals Sylvanus Vivian and George North respectively, envisaged the register as fulfilling a much wider remit and both assiduously lobbied to have the register made a permanent fact of British life. Vivian drew up plans for the register before the outbreak of war and only reluctantly held these in abeyance until the end of September 1939 when the plan was rolled out across the country. He retired in 1945, but his successor picked up the baton and argued across Whitehall for the register to be retained as what he called a "common service instrument," a central population data bank, on which all departments could draw for details of the British people as individuals, or as an aggregate.[4]

However, such post-war arguments failed and, just as Vivian had been unable to get the register introduced before hostilities commenced, so North could not persuade the government to keep it after they had finished. There are two explanations that are usually proffered to explain this abolition. The first is Lord Goddard's 1951 ruling in the case of *Willcock v Muckle*. This centred on the refusal of a member of the public, Clarence Willcock, to show an identity card to a police officer, PC Muckle. Goddard ruled in Willcock's favour arguing that it was "wholly unreasonable" for the police "to demand a national registration card from all and sundry" when the basis for doing so was an Act of Parliament that had been passed to meet an emergency that no longer existed. To continue with this scheme was, his judgement held, to turn "law-abiding subjects into lawbreakers, which is a most undesirable state of affairs."[5] The second reason given to explain the abolition of the National Register is that it was terminated as part of the government's post-war austerity campaign.

Clarence Willcock's David and Goliath-type stand adds a dash of piquancy to the story of the abolition of the National Register and his courtroom triumph is usually portrayed as sounding the death knell of registration.[6] But this ignores two things. First, that the continuation of the National Register in peacetime had already been discussed across Whitehall and that there was clearly a strong tide of opinion running against keeping it even before Willcock encountered Constable Muckle. Second, it fails to explain the government's attitude when faced with Goddard's judgement. Had the government really wanted to keep the register it could have brought its ideas before Parliament and so maintained this wartime institution in peacetime, safe from the reaches of stalwart defenders of British traditions. Similarly, the argument that registration was abolished as part of the post-war austerity drive fails to account for why the government chose to place the register under the cost-cutting axe, a decision, which it is argued here, can only be understood by examining the attitudes prevalent at the time towards population data.

As home secretaries Blunkett and Howard would be members of governments that had wholeheartedly embraced population data as a vital necessity for administration. However, their predecessor in this office immediately after the war, the Labour home secretary James Chuter Ede, advanced an argument that was diametrically opposed to his successors' and sided with Goddard in claiming that the National Register was "alien" to the British way of life.[7] Moreover, officials serving the Attlee government were also in favour of abolition maintaining that the all the data, which George North was at pains to show could be supplied by the register, was unnecessary. That they placed, at best, a low value on detailed population statistics can be seen in their conclusion that these could be obtained "if need be" from the census or the electoral register.[8]

Though Vivian and North's arguments would be resurrected more or less intact by generations of later officials and ministers, they lost the argument in this earlier period. One of the most often repeated terms used by their successors as standard bearers in the cause of government population data was the need to "modernise" the state and its practices. The usage of this term directs attention to the way they wanted the state to be and, juxtaposed to this, their view of the way the state actually was in this post-war period. Foucault defines a modern state by what it does, it is "a practice," the sum of the methods it uses to solve the problems posed by its positioning of the population as the centre of its policies.[9] These practices privilege the collection of knowledge about the population. Indeed, in his presentation of the working of these power systems, through the metaphorical device of the Panopticon prison, he describes this as "an endless extraction of knowledge" from the population. This, he argues, gives government "the measure of its quantity, mortality, natality; reckoning of the different categories of individuals in a state and of their wealth," amongst other things.[10] But having thus delineated the bounds of a modern state, Foucault was at pains to contrast the system he described and analysed in mainland Europe to "a much looser Anglicized system."[11]

That this looseness was characteristic of the British system in the post-war period is something that is widely accepted by historians. Thus, Edward Higgs has written that: "the British public associated state identification with wartime emergencies, and with the identification of the criminal, 'Johnnie Foreigner' and subject races." Jane Caplan notes that: "registration and identity cards have not normally been part of the [British] apparatus of administration." Additionally, Jon Agar has pointed out how the tensions that developed within Britain over the use of "Prussian" methods during the First World War were an updated version of a hostility that had traditionally focused on the bureaucratic power of the French state. The British contrasted these continental systems to their own, more localised, patterns of policing and record keeping with these "partial registers," Agar argues, being lauded for preserving "Britishness."[12] However, this does not analyse the nature of this Britishness: how it was preserved by the political attitudes, systems and actions (or inaction) of the state. It is argued here that, in a similar vein, without understanding the attitudes at the heart of the post-war system, it is impossible to perceive the nature of the change that occurred in the 1960s.

Foucault positions the "human sciences" that analyse people as living beings and working individuals as part of the central motor force of the systems of knowledge that characterise modernity. He argues that this knowledge is reciprocally linked with power thus: "police and statistics mutually condition each other." However, elsewhere he added a caveat to this generalised view that is important for the point made here, thus: "I think the meaning of the English word 'police' is something very different."[13] Foucault is clear that though there is a reciprocity linking knowledge of the population and power over it, that the power to initiate this relationship lays with the state: "police makes statistics necessary, but police also makes statistics possible."[14] Theda Skocpol concurs, noting that: "governments and their activities have profoundly affected the emergence and social organization of social science disciplines, as well as their intellectual orientations," to which general point she adds that the nature of the government and its activities led to "the retardation of the professionalization of the academic social sciences in Britain."[15] Thus, it is here, in the political attitudes that produced only a tenuous linkage between the knowledge of the human sciences, social science and medicine, on the one hand, and the British government on the other, that the explanation for the post-war abolition of the National Register in particular, and government attitudes to data about the population in general, should be located.

2

The attitudes that characterised the political traditions of Britishness, as this related to population data, were encapsulated by the chancellor, Sir John Anderson, in October 1944, when he wrote to Clement Attlee, Labour leader and deputy prime minister in the wartime coalition, referring to "the so-called social sciences."[16] Though Attlee did not have such a dismissive attitude, it is, nevertheless, the case that his post-war government limited its actions to establishing committees to inquire whether there was a need for social science research and, if there were, what might be done to aid it. These post-war committees thus give valuable insights into the mind of the British government and the place of social science in Britain. Harold Wilson would later describe data as "the sinews of all social science" and an examination of the work of these committees shows that, for as long as the government only pursued data in a fragmented and ad-hoc manner, or refused to share what data it held with anyone outside the civil service, social science would be hamstrung through being starved of the means to establish itself at the heart of British politics.[17]

That this lack of data severely hampered both social science and government was a point forcefully made, in February 1946, when William Hancock (Professor of economic history at Oxford and editor of the civil service history of the Second World War) gave evidence to the Committee on Social and Economic Research (the Clapham Committee: it was set up in 1945 and charged with looking into the financing of social research). He argued that the war had exposed the government's "considerable lack of knowledge as regards social matters." Moreover, he maintained that the information that did exist within government circles was severely constrained in scope with, for example, the knowledge possessed by the Ministry of Health being limited only to the narrowest definition of health, with the result that it failed to include any social and economic data. This, he explained, had exposed the wartime government's almost total lack of understanding of, for example, how "valid" an institution the family actually was to the life of ordinary British people.[18] In the committee's discussion of Hancock's evidence, the point was made that one reason for this narrowness of vision was that the government's information was held separately by different departments that did not share it with others across Whitehall. Hancock's suggested remedy for this was to bring in outside researchers who would have a greater breadth of vision and understanding, while others on the committee proposed the establishment of a social science research organisation to pursue more or less the same aims.

However, what is also made clear from this post-war work is the extent to which being starved of the material on which to conduct research had severely cramped the development of the social sciences. Clapham's committee began its life by receiving a copy of a 1945 report on the co-ordination of the social sciences. This had been produced by the Consultative Conference on Economic and Social Research that had been set up at the start of the war to coordinate the distribution of research grants from the Rockefeller Foundation. This report argued that the social sciences did not need a national body to represent them for the simple reason that, given "the present state of the Social Sciences in this country," there was very little to actually represent.[19] What existed were discreet academic disciplines that were so different from one another that the generic notion of social science was wholly inappropriate. Moreover, these disciplines "vary widely in the degree to which they have worked out an adequate technique of study and secured authority for their results."[20] On this basis, the report argued that what was needed was for serious academics engaged in research to carry on building departments in universities and so develop a critical mass of social science in Britain.

There were others who agreed with this conclusion that academia and politics should keep each other at arm's length as a point of principle. Sir John Anderson, for example, maintained that government should not work with social scientists because doing so would impinge on the academic freedom of universities to pursue "fundamental knowledge."[21] Similarly, when the cabinet secretary, Burke Trend, wrote a report on the funding of science research, he argued that control of the research agenda was not a proper function of government. Moreover, and more importantly, this report had nothing at all to say about the social sciences.[22] But the social scientists on the Clapham Committee did not argue that they should refuse to deal with the government at all, but rather that they were not strong enough to do so. Thus, Thomas Simey (Professor of Social Science at Liverpool University and, from 1965, a life peer) told the Clapham Committee that his university colleagues did not think that his field of sociology was "altogether respectable." However, he added that he could not really complain about this "since sociologists had not yet done the work which they ought to have done to make theirs a fundamental study."[23] All of which was seen in the fact that there were only two professorial chairs in sociology in the country (Simey's and another in London). The Clapham Committee's report dug further into this matter and discovered that while there were a total of 889 professors in Britain (in 1938–39) only thirty-five pursued social sciences. Moreover, these thirty-five chairs were divided amongst people studying eleven different disciplines whereas, by contrast, the nineteen professors of agriculture were divided into only three separate disciplines.[24] The Clapham Report did not pull its punches on this score stating plainly that when it came to any form of research into social questions, and regardless of what facet of such research work was examined, British universities were "lamentably inadequate."[25] Although it insisted that this damning indictment applied across all the social sciences, the report singled out the study of statistics for special attention. During the war, the report argued, "statisticians were probably the scarcest of scarce commodities." However, at the time of writing their report, the committee noted that the situation had actually deteriorated since, "the demands which are certain to accrue as current policies develop and projected institutions are brought into being," could never be met by universities that held fewer chairs in statistics than "could be counted on the fingers of one hand."[26]

Simey referred to the social sciences as being caught in a "vicious circle" and Clapham similarly argued that they were starved of both the data they needed to conduct research and of the opportunities to work on government projects.[27] Moreover, with regard to both of these, the report was clear that "the blame must rest squarely upon governments."[28] The social sciences needed population data, but they were unable to collect this to anything like the extent that the government could. Moreover, government departments held stores of information that they collected during the discharge of routine administration that could be a rich seam of information about the population to be mined by researchers with, "much to be gained on both sides from the contacts which it would involve."[29] But this cooperation was not usually forthcoming. Simey told the committee that before the war, the Unemployment Assistance Board in Liverpool had allowed his students to attend interviews with claimants. This work had led to a report being given to the authorities advising on their handling of "difficult cases" and these recommendations had been gratefully received. However, since the war, this cooperation had been stopped "on instruction from London."[30] If, as Foucault noted, the British system was markedly different to that in mainland Europe, this was thus because, as Simey (and Clapham) noted, the central motor of the government was not driving a system of population data in Britain the way it did across much of the rest of the continent.

Having surveyed the weaknesses of British social science, Clapham did not accept that a social and economic research council, along the lines of the existing Medical Research Council, would solve anything, because as Clapham defined it, the problem was not how to allocate resources to research, rather it was how to get the data to conduct research on. As an example of the weakness of British statistics, the Clapham Committee, fully aware of the irony of this, had given the statistics (above) of the number of social science professors in Britain, which at the time though seven years old, were nevertheless the most up-to-date information available.[31] Having thus determined that the chronic lack of contact and communication between social scientists and government was acting to the detriment of both parties, Clapham recommended the establishment of an interdepartmental committee. This would be "charged with the duty of bringing to the notice of departments the potential value for research purposes of the material which they collect and suggesting new methods and areas of collection."[32] A month after the committee reported this new body was established. George North, the registrar general, who Sir Edward Bridges (the head of the civil service) was told, was "much interested in the question," became its chair.[33]

The North Committee agreed with Clapham that only government could provide the data that social science needed. The committee thus saw its raison d'être as being to nurture the latent symbiosis between government and social science by helping government departments to facilitate the use of their data by outside researchers.[34] This was a process that had begun during the war and in both its published reports, the committee argued that this process needed to increase in both depth and breadth. It maintained that such a post-war increase was necessary because an increasing number of aspects of peoples' lives were coming under the aegis of government and that this had two effects on the nascent relationship between government and social science. First, government departments had a heightened need to analyse information to underpin this increasing field of policy and clearly needed help in doing so. Second, because "changes in the administration of such services as those connected with public assistance, with the hospitals and with social insurance," meant that "information formerly collected by local authorities or private agencies is now gathered into Departmental files."[35] This increased the government's need for help in understanding what this enlarged stash of information actually meant, while simultaneously making this potentially centralised data cache all the more interesting to the social scientists on the committee.

In order to achieve this strategic aim, the North Committee set out to investigate what information departments actually held; whether any of this could, or should, be published and to make any recommendations that it could for changes to the gathering of statistical data by departments. As the committee noted, even the first of these was "a considerable task" in itself partly because, as William Hancock had told them, "valuable data are apt to be found widely dispersed among a vast mass of papers of which many are inadequately indexed."[36]

The committee decided to begin its trawl at the Ministry of Labour and National Service where it made recommendations about the collection and publication of reports into absenteeism, working hours and family budgeting.[37] It also applied itself to the Board of Trade and the Ministry of Education. At the former, the committee made a series of recommendations about how reformed data, including details of newly registered patents, could be published in the *Board of Trade Journal.* The Ministry of Education had, under the terms of the 1944 Education Act, been given for the first time the ability to conduct research and the committee took a keen interest in this and praised the early steps taken here. The committee had thought that the ministry could use its records to compile cohort studies of children as they passed through the school system in order to evaluate the factors that had an impact on their education. However, given that the system was in a state of considerable flux, it thought that such studies should be delayed. Nevertheless, it did recommend that local authorities should keep detailed records of all the children in their areas since this would be a valuable resource for future researchers.[38] In addition to the work of these subcommittees, assigned to particular departments, the committee also made suggestions for how the Public Record Office might be reconfigured to aid research into contemporary society.[39]

One area where the committee was keen to establish that it had made a real impact was in the newly introduced census of distribution. In 1948, the committee reported that it was involved in analysing pilots for this survey, and in 1950, it held the census up as an example of the new types of action taken by government that required the assistance of social scientists.

However, events around the census of distribution reveal that the committee was seriously misreading the government's attitude towards data in post-war Britain and that in doing so, it was much too optimistic about the post-war system's willingness or ability to accept change.

The census of distribution was set in motion just before the Labour Party took power by a minute issued in June 1945. This established a committee that reported in March 1946 and, in April, the census planning committee began work on the pilot census.[40] In 1947, the Statistics of Trade Act was passed and arrangements were set in place to take the census in 1948. However, this was postponed and rescheduled for 1949, but when the time came, it was again postponed and was not actually carried out until 1950 being published in 1952. Just after this second delay, Hugh Dalton, who was chair of the Cabinet Committee on Distribution and Marketing, wrote to the prime minister, Attlee, about the "desultory" pace of the civil service's work on the census adding: "I have tried to infuse them with a sense of urgency. But I have become impatiently conscious of a desperate disinclination to decide, transmitted upwards by officials in more than one department."[41]

This lack of enthusiasm for data collection was not confined to those officials working on the census of distribution. Indeed, shortly after the North committee published its first report, at the time, when the committee began to approach the Board of Trade, opinion in that department coalesced around proposals to defend their existing patterns of behaviour that thus pre-empted North's suggestion by buttressing the department's systems. Thus, while seeking to avoid a formal meeting, guidance was circulated warning that: "anything which we may say about our practice on this subject may eventually be published and lead to both criticism of our policy and to embarrassing requests for access to particular classes of information."[42] In addressing this issue, the guidance made three points. First, that though the board's papers could be used by its officials even here, if the product of such research were to be circulated across Whitehall, great care would be needed in its use. Second, it was argued that if researchers from other departments were to use the board's material, there would need to be strict guidelines to prevent "uncontrolled dissemination or public disclosure." Third, the guidance maintained that researchers from outside the civil service would not be given access to official papers except in "exceptional circumstances" and, even then, only if they were subject to a departmental veto over the end-use of such material. In dealing with less exceptional requests for access, the normal response, officials were instructed, should be to offer the potential researcher only a general verbal discussion of the nature of the material held by the board.[43] Any decision concerning access would reside solely with the board, there would be no court of appeal and, as a guiding principle, these notes suggested that: "there should be absolutely no question of any direct access to this material by non-official outside researchers."[44]

Indeed, so loathe were officials across Whitehall to enhance their own data-gathering operations, or to use social scientists to analyse these holdings that, though it soldiered on for several years, the North Committee was pushed further into the Whitehall sidings than ever. Thus, after the Conservatives returned to power in October 1951, though the committee's third report was drafted, it was never actually published. What this report shows is a body that was aware of its own impotence. The report noted that at its inception, the committee addressed questions of "broad general import," whereas, by 1955, it rarely met as a whole group and focused instead on what were called "more specific and practical matters."[45] In this last report, the committee drew up a list of the outcomes of its recommendations, or rather a list of what it thought these results might be since it had no real way of knowing whether these had been enacted. This was the case because it had never possessed any means of obliging departments to follow them. Indeed, the report noted that its work "was not sufficiently recognised" along Whitehall.[46] This slide into insignificance is all of a piece with the general pan-Whitehall attitude with regard to research. Moreover, such attitudes within the civil service were not only confined to Whitehall's relationships with social scientists.

3

One of the main, if not indeed the main, area of British life that came increasingly under government control as policies developed and projected institutions came into being in the post-war period was, of course, medicine. Like social scientists, medical researchers would become an important source of pressure for access to government records. But more than this, medical researchers would staunchly advocate the transformation of medical data into a standardised, centralised and linked system. In 1963, medical researchers in the Oxford area led by Donald Acheson (he would be UK chief medical officer, 1983–91), began an experiment that centred on reorganising all medical records to find out whether they could create "a single integrated file of health data for a community."[47] That this experiment needed to be conducted at all, fifteen years after the creation of the National Health Service (NHS), indicates that though the government may have taken control of the politics and finances of the health system, it was not, at this point, operating this as a biopolitical data system and was still dominated by the same set of attitudes that led to the abolition of the war-time register and which spoke of "so-called" social science. In fact, Acheson described the system he had to work with as "absurd," "flaccid and disorganised" and subject to "arbitrary fragmentation." He noted that a previous researcher had commented that the detached nature of its components meant that the system was akin to "an animal with bones, but without a spinal column." However, he took this analogy further to suggest that it might be more appropriate to regard the elements of the health-data system "as ganglia each with their ... nerves but without connections. An animal with such a nervous system would be classified as primitive, and would have little chance of co-ordinated or intelligent behaviour."[48] This primitive and fragmentary "avalanche of paper" could function to assist medical treatment only as long as such treatment were confined to helping patients who presented with problems.[49] For preventive medicine it was useless, and it was wholly incapable of meeting the needs of any coordinated system of epidemiological research. In short, it was not a system at all.

In the long term Acheson saw computerisation as the best method of resolving this series of problems and he became a great advocate of the new technology, earning himself a reputation for promoting it with an "unrealistic zeal" across Whitehall.[50] But, in the short term, he also sought to standardise the way in which data was collected and collated inside the NHS. To this end, he welcomed the way in which the 1965 Tunbridge Report on the Standardisation of Medical Records argued for clinical records to be made available nationally for use by medical researchers. In particular, he wholly endorsed the report's drive to ensure that records were identifiable, rather than issued as anonymous statistical aggregates as this would, for the first time, allow hospital data to be related to other sets of records. However, he argued that the system proposed would have benefitted from a pilot study, and in this, he seems to have been correct. Thus, a departmental report of 1973 recorded that after 1965, only about one-third of British hospitals used the new standardised forms. A study undertaken in 1969 to get to the bottom of this found that though hospitals supported standardisation they were hostile to the format of new forms. These were redesigned and the ministry strongly urged hospitals to use them (having rejected a proposal to make this adoption mandatory) in 1972 and reiterated this advice in 1973.[51]

Acheson though was far less concerned with this sort of detail than he was with the big picture of NHS data flows. Operating at the forefront of medical research, he maintained that: "reorganisation involving the development of sound lines of communication and much more beside must come first." Moreover, he fully recognised the overarching scale of what he was proposing.[52] Thus, he noted that building such communication architecture would involve the "rationalisation of the structure of local government and [the] resolution of conflicting boundaries."[53] Indeed, his strategic aim was not to make sure that the NHS had a data system, rather he sought to make the NHS into a data system. He was also clear that though he worked on health data, the system he envisaged "may be applied to any field in which it is necessary to bring together information recorded about persons in different places or at different times."[54] Indeed, his system was predicated on doing just that, not only for health data, but also a wide range of data that was not medical, or which was administrative in nature and had not been collected for research of any kind at all. Thus, he argued that not only should the numbering system of the Ministry of Social Security be merged with that of the NHS, but that the entire filing systems of both should become one.[55]

Acheson was not a lone voice crying in the wilderness, his were ideas that were heard and acted upon. In 1964, he presented a long memorandum based on his Oxford study to the Medical Research Council's (MRC) Committee on General Epidemiology, which was influential in the drafting of the committee's 1965 report examining proposals for a national system of linked medical records.[56] This is important because not only does it demonstrate the direction of medical thinking, it also provided the government with a concrete proposal around which its own thinking could coalesce. This MRC plan arrived at the General Register Office at a time when the Registrar General, Michael Reed, was one of those working with the new Labour government on ways to reintroduce the wartime population register and he saw the MRC's plans as an opportunity "to carry our own thinking on record linkage a good deal further than we have been able."[57]

In wanting to unite the records of the Ministry of Health with those of the Ministry of Social Security, Acheson can be seen as representing a broader current in post-war thinking that sought to connect social and medical matters, if not indeed merge them into one epistemological, political and administrative whole. Thus, in 1945, the MRC passed a resolution noting that "the separation of research in vital statistics and industrial psychology from medical research is to be deprecated."[58] While a few months later, as has been seen, William Hancock argued that it was a mistake to separate health and socio-economic data. This was also the underpinning for the establishment of the Social Medical Unit (SMU, it was later renamed the Social Medicine Research Unit). This was set up at the beginning of 1948 and was the product of lobbying of the MRC by the epidemiologist Professor Jerry Morris (the SMU's head until it closed in 1975) and his good friend and colleague Richard Titmuss (he was briefly the unit's statistician). Titmuss would later decry Britain's "appalling lack of facts" as "irresponsible" and the two men were determined to rectify this shortfall.[59] In their proposal to set up the SMU, they stated that the main principle behind their work would be "the marriage of vital statistics with local social and medical enquiries" and argued that through this union their work would develop "the scope and value of the vital statistics themselves."[60] The paucity of the data available was, they argued, witnessed by the Ministry of Labour's cost of living survey. This had not been conducted since before the war; however, even at that point, it could not have represented the reality of life for working people, since it was based on models developed in 1906.[61] To address this, the SMU proposed that it, along with the government social survey, should run a survey of the population of Willesden (where the SMU was based). This survey would be a local census that would gather qualitative data on how people felt as users of the NHS, along with a longitudinal study to assist the NHS in planning its services. For everyone involved with the SMU, medicine was obviously a social science and by the 1960s, the unit's work had taken a markedly sociological turn with investigations into juvenile delinquency and educational attainment.[62]

The SMU did not only argue that Britain lacked population data but rather, like Acheson, Morris maintained that such information as did exist was fractured across different ministries, thus:

> a system of identification needs to be established and built into the official administrative system … [to] enable the linking together of extracts, as required for specific research, from the hundreds of millions of entries now being made in the NHS etc.[63]

Additionally, like Acheson, he wanted such a system to be designed and deployed by the MRC. Thus, above all else, what both men sought was, access to data and both found their work stymied by a paucity of information, a lack of access to it or its fragmentary nature. Their advocacy of an integrated system to be built and run by the MRC was thus in many senses the product of their despair at what the government held, or was prepared to put at their disposal. But their planning and proposals were an integral part of the pressure that mounted through the post-war period and which, perhaps paradoxically, would make their efforts redundant. This was the case because in 1964, a Labour government was elected that shared their views of the importance of population data. This government would thus set about conducting research to gather information on the population and, most importantly, would seek to construct just the data matrix that Acheson, Morris, Titmuss, Hancock and Simey, had argued for throughout the post-war period. It is in this new intellectual milieu, a radical departure built in opposition to the looseness of traditional British political culture, that the nexus of the British state's modernising data turn should be located.

Notes

1 *Identity Cards: A Summary of Findings from the Consultation Exercise on Entitlement Cards and Identity Fraud*, Cm. 6019 (Nov. 2003), 21.
2 *Identity Cards: A Consultation Document*, Cm. 2879 (May 1995), 6.
3 See Kevin Manton, *Population Registers and Privacy in Britain, 1936–1984* (London: Palgrave MacMillan, 2019).
4 TNA, T 222/436, George North, *National Registration*, 19 Apr. 1951, 2.
5 Lord Chief Justice Goddard, 26 June 1951, in, *Select Committee on Home Affairs Fourth Report*, 30 July 2004, 1–2.
6 See, for example Benjamin J. Goold and Daniel Neyland, "Introduction: Where Next for Surveillance Studies," in *New Directions in Surveillance and Privacy*, eds. Benjamin J. Goold and Daniel Neyland (Uffculme: Willan, 2009), xix; 105; Mark Egan, "Harry Willcock the Forgotten Champion of Liberalism," *Journal of Liberal Democrat History* 17 (1997–98): 16.
7 TNA, RG 28/300, quoted in, E. H., *The National Register and the Identity Card*, 21 July 1950, 1.
8 TNA, WO 32/21649, *Report of the Committee on National Registration*, Feb. 1950, 4.
9 Michel Foucault, *Security, Territory, Population* (New York: Picador, 2004), 277.
10 Michel Foucault, *Psychiatric Power* (New York: Picador, 2006), 77; and Foucault, *Security, Territory, Population*, 274.

11 Michel Foucault, *The Birth of the Clinic* (London: Routledge, 1989), 48.
12 Edward Higgs, "Consuming Identity and Consuming the State in Britain Since c.1750," in *Identification and Registration Practices in Transnational Perspective: People, Papers and Practices*, eds. Ilsen About, James Brown and Gayle Lonergan (Basingstoke: Palgrave, 2013), 177; Jane Caplan, "'This or That Particular Person': Protocols of Identification in Nineteenth-Century Europe," in *Documenting Individual Identity: The Development of State Practices in the Modern World*, eds. Jane Caplan and John Torpey (Princeton: Princeton University Press, 2001), 64; Jon Agar, *The Government Machine: A Revolutionary History of the Computer* (London: Massachusetts Institute of Technology Press, 2003), 137.
13 Foucault, *Security, Territory, Population*, 317, and, Michel Foucault, *Power: Essential Works of Michel Foucault 1954–1984*, ed. James D. Faubion (London: Penguin, 2002), 410.
14 Foucault, *Security, Territory, Population*, 315.
15 Theda Skocpol, "Governmental Structures, Social Science and Politics," in *Social Science Research and Government: Comparative Essays on Britain and the United States*, ed. Martin Bulmer (Cambridge: CUP, 1987), 42 and 44.
16 TNA, T 161/1301, Anderson to Attlee, 10 Oct. 1944, 1.
17 TNA, CAB 139/742, *Draft Speech for the Royal Society of Statisticians Banquet on the Occasion of the 37th Session of the International Statistical Institute*, September 10, 1969, 10.
18 TNA, T 161/1301, *Committee on Social and Economic Research, Meeting 11 Feb. 1946*, 1.
19 TNA, T 161/1301, *Report on Co-ordination of Social Sciences*, 7, in, Carr-Saunders to Clapham, 7 Aug. 1945.
20 Ibid., 3.
21 TNA, T 161/1301, Anderson to Attlee, 10 Oct. 1944, 1.
22 *Report of the Committee of Enquiry into the Organisation of Civil Science*, Cmnd. 2171 (1963).
23 TNA, T 161/1301, *Committee on Social and Economic Research, Minutes of a Meeting*, 24 Oct. 1945, 1.
24 *Report of the Committee on the Provision for Social and Economic Research*, Cmd. 6868 (July 1946), 7–8 and 14.
25 Ibid., 9.
26 Ibid., 8–9.
27 TNA, T 161/1301, *Committee on Social and Economic Research*, 24 Oct. 1945, 2.
28 *Report of the Committee on the Provision for Social and Economic Research*, 5.
29 Ibid., 6.
30 TNA, T 161/1301, *Committee on Social and Economic Research*, 24 Oct. *1945*, 2.
31 *Report of the Committee on the Provision for Social and Economic Research*, 7.
32 Ibid., 6.
33 TNA, T 161/1301, to Sir Edward Bridges, 8 Aug. 1946, 2.
34 *Report of the Interdepartmental Committee on Social and Economic Research*, Cmd. 7537 (Oct. 1948), 3 and 4, and, *Report of the Interdepartmental Committee on Social and Economic Research*, Cmd. 8091 (Dec. 1950), 1.
35 *Report of the Interdepartmental Committee*, (1950), 1.
36 *Report of the Interdepartmental Committee*, (1948), 9.
37 *Ibid.*, 7.
38 Ibid., 7 & 8, and, *Report of the Interdepartmental Committee*, (1950), 7.
39 *Report of the Interdepartmental Committee*, (1948) 7 and 8, and, *Report of the Interdepartmental Committee*, (1950), 5 and 6.
40 TNA, BT 70/121, Census of Distribution, Planning Committee, *Minutes*, 1946.

41 TNA, PREM 8/1420, Memo from Dalton to Attlee, '*Committee on Distribution and Marketing*', 27 July 1950.
42 TNA, BT 64/1349, *Access to Confidential Information*, 10 Nov. 1948, 1.
43 Ibid., 2.
44 Ibid., 3.
45 TNA, RG 25/8, *The North Committee, Third Report, Draft*, 8 Nov. 1955, 2.
46 Ibid., 7.
47 E. Donald.Acheson and John Grimley Evans, "The Oxford Record Linkage Study: A Review of the Method with some Preliminary Results," *Proceedings of the Royal Society of Medicine* 57 (1964): 11.
48 E. Donald Acheson, *Medical Record Linkage* (London: OUP and the Nuffield Provincial Hospitals Trust, 1967), 135, 183, 2, and 52.
49 Ibid., 11.
50 TNA, RG 48/149, *Record Linkage*, 24 June 1966, 2.
51 TNA, MH 160/1180, *Standardisation of Hospital Medical Records*, n.d., ca. June 1973.
52 Acheson, *Medical Record Linkage*, 183.
53 Ibid., 53.
54 Ibid., 1.
55 Ibid., 89, 106, and 156.
56 TNA, RG 48/3149, E. Donald Acheson, *Proposals for the Reorganisation and Integration of Medical Information into a Single National System*, 14 Sept. 1964; TNA, FD 1/8310, "Medical Research Council, Committee on General Epidemiology, Sub-committee on Record Linkage," *Proposals for a National System of Linked Medical Records*, Aug. 1965.
57 TNA, RG 48/3149, R.G.O., *Record Linkage*, 13 Aug. 1965, 1.
58 TNA, T 161/1301, Privy Council Office to T.M. Wilson, 3 Mar. 1945.
59 Richard M. Titmuss, *The Irresponsible Society* (London: Fabian Tract 323), 8.
60 TNA, FD 1/284, Jerry. N. Morris and Richard. M. Titmuss, *Proposed Social Medical Research Unit*, 11 Feb. 1947, 1.
61 TNA, FD 1/284, Richard M. Titmuss, *Parenthood and Social Change*, 15.
62 Ann Oakley, "Fifty years of J.N. Morris's Uses of Epidemiology," *International Journal of Epidemiology* 36 (2007): 1185; and, TNA, FD 12/522, "Social Medicine Research Unit," *Progress Report, 1964–68*, and Social Medicine Unit, *Progress Report, 1968–72*.
63 TNA, FD 12/1637, Jerry N. Morris, *Review of the Field of Social Medicine, Background Notes for a Talk to be Delivered to the MRC on 19 April, 1963*, 9 Apr. 1963, 3.

2 Government 1964–79 and the Growth of the Modern Data State

The previous chapter sketched the general outline of a logjam in British population data caused by a combination of the government's unwillingness to share such data as it held with outsiders, and its inability to turn its data holdings into a recognisably modern, linked system. It indicated how this lack of access restricted the development of those, such as social scientists and medical researchers, who would have thrived on government data stockpiles while, simultaneously, restricting the ways in which government could use its data thus constraining the ways in which it could think and act. By the end of the period covered by that chapter Peter Townsend (Professor of International Social Policy, Chair of the Fabian Society 1965–66 and, in 1965, founder of the Child Poverty Action Group) was one of those applying pressure on the National Assistance Board to change this system. In 1964, he was one of the doctoral supervisors of a research student, Dennis Marsden, who approached the board for access to data about "fatherless families." In accordance with the traditional Whitehall attitude on confidentiality, this researcher was refused any such access and "when he realised he would not get it, he commented," that his supervisors "would probably decide to wait for a change of government and then try again."[1] Townsend's understanding of the way British politics was moving was prescient. Indeed, in 1966, eighteen months after the election that would usher Labour into power, the National Assistance Board noted that: "the general drift of opinion and events is towards freer co-operation between government departments and research workers," and that this drift was "a result of increasing pressure which has been brought upon the Department in the last year."[2] In other words, the pressure for change had begun when Labour, under Harold Wilson, took office and opened the gates to encourage this new current of thinking to flow along Whitehall.

Wilson was member of the Royal Statistical Society (RSS) for twenty-six years and served on the society's council for the two years up to his election as prime minister. When he made a speech at a banquet held by the society in September 1969, he was clearly proud to be able to tell those assembled that: "in statistical organisation and techniques we are passing through a revolutionary period in the development of the Government's Statistical

DOI: 10.4324/9781003252504-3

Services."[3] This chapter aims to outline the contours of this revolutionary data programme, and to do this, it is divided into three sections. The first of these examines what the government's rhetoric of modernisation meant in population data terms. It explains how breaking the logjam that restrained biopolitics in Britain put government on the trail of the chimera of perfect data that thus initiated many of the subsequent developments covered by this book. In the short term, this led to the Wilson government proposing the reintroduction of a registration system similar to the one that had been abolished in 1952. The government was unable to secure its data revolution in this way and the second section details how this strategic setback, which left it unable to outflank the constraints inherent in the existing population data system, put it on course to engage in a war of position against codes of confidentiality and the power of professionals within this traditional system. The third section addresses the final big issue facing any British government that wanted to revolutionise its data operations. This was the fact that not only were stocks of data ring-fenced from each other by codes of confidentiality but, the institutions that gathered data were also highly decentralised. Any government wanting to unite its data holdings would thus have to be a centralising government, and this would involve it in yet further political struggles against the status quo.

1

Reflecting on his premiership, Wilson stated that the key moment in establishing a flow of population data was the *Fourth Report of the Estimates Committee* of December 1966.[4] However, the fact that this report, authorising spending and allowing reform to go ahead, was commissioned and so warmly received by the government reveals that the real political impetus behind this data turn came from the top of a new administration which recognised that Britain was "in a period of rapid social change" and was determined to understand this in order to steer it.[5] Indeed, less than two months after taking office, Wilson received a report on *Economic and Social Intelligence*, which set out his government's highly critical view of the population data streams available to it. This detailed how for many fields of policy there was, at best, only a scant supply of population data and that even when this was available, it was often out of date or inaccurate.[6] In other words, this report, and the ministers reading it, agreed with Acheson, Morris and Titmuss that what existed was so "arbitrary," "fragmented," "flaccid" and "disorganised" that it was barely a system at all.[7] This view was reinforced when, in 1965, the government received the report of the Heyworth Committee on Social Studies. This report argued that "much still remains to be done in respect of the provision, co-ordination and publication of social statistics which involves relationships both among Government departments and social scientists in universities and other research establishments."[8] It suggested that the best body to achieve this increased coordination would be the Central Statistical Office (CSO) that, it argued, ought to provide the chair for a committee with the ability to cut across Whitehall departmental boundaries. Additionally, the report successfully made the case for the establishment of a British social science research council that would be responsible for furthering the inclusion of outside researchers. During the Second World War, the government had established the social survey. This was still running in the mid-1960s but Heyworth maintained that the way the survey was organised prevented it from acting effectively. The problem was, he argued, that the survey had no knowledge of the policies being developed inside departments, no control over the demands placed upon it and no way of reconciling conflicts that might arise over the use of its skills. To resolve this set of issues, Heyworth argued that the survey needed to be relocated at the centre of Whitehall power, and he recommended that it be run through a committee chaired by the Treasury. Were this done, he maintained, the survey would be much better placed to affect the training of researchers, and "the accumulation of a coherent body of knowledge as a background of Government administration."[9] This desire to build a body of experts and knowledge revealed that the attitudes behind the data turn were anything but a flash in the pan.

Thus, in 1968, the government received a report from a committee, of which Jerry Morris (see Chapter 1) was a member. This, the Seebohm Committee, enquired into Local Authority and Allied Personal Services. Its report damned the fragmentation of government data about services provided to the elderly and disabled, arguing that this prevented government from knowing the impact of its policies: "we should by now have better answers to such questions."[10] The following year, at his speech to the RSS's banquet (above), Wilson told his audience that: "social statistics … have been relatively starved of resources."[11] Moreover, this was a view that was echoed both a year later by the Ministerial Committee of Social Services and a year earlier when, at the government's Statistical Policy Committee, it was concluded that: "it is clear that, until recently the importance of statistics was not fully recognised in the government service."[12] All of these, characteristic comments, might be seen as the zeal of enthusiasts given free rein to pursue their pet project. But what needs to be noted here is that these comments, about the paucity of the government's data resources, did not stop after a few years. Moreover, this government attitude continued after Wilson lost the 1970 election. Thus in 1972, the chief statistical officer, Claus Moser (he held this post from 1967 to 1978), chaired a meeting that was set up to encourage departments to provide data that would allow the development of knowledge about the background of social-service clients.[13] In fact, this attitude, ever critical of the extent of data provision and always in pursuit of more, became part of the warp and weave of the British government. For example, in the early 1980s, reference was made to "database horror stories" while at the start of the following decade, Philip Redfern (deputy director of the Office of Population Censuses and Surveys [OPCS] 1970–82) described the government's population data system as lax, uncoordinated, ramshackle and riddled with "incredible" missed opportunities.[14] The same attitude was evidenced when, in 2005, the prime minister, Tony Blair, criticised government systems that he characterised as "islands with their own data, infrastructure and security and identity procedures." These, he argued, made it "difficult to work with other parts of government or the voluntary and community sector."[15]

Reading Blair's words might encourage the belief that nothing at all had changed in the thirty-eight years that had passed since Acheson had excoriated the primitive and uncoordinated collections of paper that constituted British population data. However, the reality was that the government's data systems, as is shown throughout this book, expanded markedly after the mid-1960s. Indeed, some of the remarks cited above were made in committees that had been produced by the Wilson government's changes to the state's data architecture and which drove the government's agenda forward. Thus, rather than describing the reality of the situation they found themselves in, these expressions of exasperation speak of the nature of the task these people were seeking to undertake and the scale of the change that this represented when seen next to the attitudes that had held sway previously. In defending his own pursuit of data about the population Redfern once noted that: "to my mind anonymity is a fantasy."[16] But the all-encompassing knowledge of the population that he and others sought was equally chimerical and this was why, as another OPCS statistician had it: "we are always looking for fresh sources of information."[17]

Foucault has established how this pursuit of the idea of population, which "directed knowledge to the sciences of life, labour and production"[18] was not a positivist search for a knowable, objectively verifiable reality. Rather it was "a constant interplay between techniques of power and their object [that] carves out in reality, as a field of reality, population and its specific phenomena."[19] Thus, the reciprocity between the search for knowledge and the objects studied, the people, does not find the population so much as create it. But precisely because the population is a product of this searching, further searching can parse this knowledge and so constantly recreate the population. The population, in other words, is "permanently and ceaselessly" variable: it is something that can never really be found or known.[20] Therefore, once knowledge framed in this way was integrated into state power, it necessarily became the permanent "discourse of a centred, and centralising power" because government was, from this point on, engaged in what amounted to a search for a chimera: destined to be "always looking … for fresh information."[21] As the Seebohm Report put it: "nor is it sufficient for research to be done spasmodically however good it be. It must be a continuing process, accepted as a familiar and permanent feature of any department or agency concerned with social provision."[22] In launching this process, Wilson had not just started a revolution; he had initiated a permanent revolution. Moreover, he was aware of this. Thus, in addressing the

RSS, in 1972, he noted that though the impetus of the Fulton reforms of the civil service seemed to be "running out of steam," the same could not be said of his reforms of the government's statistical operation since this had "developed its own momentum."[23]

The Seebohm Report was clear that social services needed to be planned to prevent problems, such as those that had recently arisen around the issue of immigration, and that this planning could not be undertaken without data. Moreover, though it argued that government needed to be the central producer of data, when it came to analysing this, "the universities and research institutes must and should be encouraged to play their essential part."[24] In fact, it was in order to facilitate such participation that the government had, as early in its term of office as 1965, followed the recommendation of the Heyworth Report and set up the Social Science Research Council. But more important than this was the change in attitude that it initiated. In the build up to the 1964 election, Professor Peter Townsend's research student had been denied access to data held by the National Assistance Board, but in 1966, the chairman of this board wrote to the minister of pensions and national insurance, Margaret (Peggy) Herbison, saying that his organisation believed: "that it was right to take the initiative" and open its data to outside researchers since "research into social problems cannot be a government monopoly."[25]

Social scientists like David Donnison (he became the second professor in Titmuss's department at the London School of Economics) saw their research as "a national resource," and the Wilson government agreed and brought them into its fold.[26] In 1965, Donnison was a member of the government's Housing Advisory Committee; he became deputy chair of the Supplementary Benefits Commission for two years from 1973 and its chair for the following five years. As was noted in Chapter 1, Thomas Simey, professor of social science at Liverpool University, was made a life peer by the Labour government in 1965. He would be joined in the upper house by six other university social scientists between 1964 and 1970. However, in an even bigger departure from the norms of the immediate post-war period, Wilson's government also agreed with social scientists like Donnison about what needed to be done to its data systems in order to make the population more legible to government. This view was reflected in, and reinforced by, the Seebohm Report. This argued that while data should continue to be collected locally, the priorities for such work, the techniques to be employed, the definition of terms used and the amount of information to be collected should all be decided nationally. Additionally, the report maintained that since this national organisation should cut across departmental boundaries to cover, for instance, both health and social security, it would have to be created anew.[27]

By this point, momentum was clearly building behind such plans for a radical re-casting of the government's data apparatus into a system that would be based on the principles that data should be standardised, centralised and linked. Chapter 1 showed how the registrar general, Michael Reed, embraced the Medical Research Council's (MRC) plans for a linked system of medical records. But he and the government would eventually take this further to fulfil the promise held out by Acheson that the system he had demonstrated, through his Oxford study of medical records, could "be applied to any field in which it is necessary to bring together information recorded about persons in different places or at different times."[28] To do this, the government would attempt, in 1969, to introduce a population register of the kind used during the Second World War and abandoned in 1952.

Plans for this were already in train in 1966 when the registrar general's office stated that schemes under consideration would, if adopted, lead to "the restoration of national registration with at least the degree of completeness which was obtained during the last war and up to 1952."[29] But these plans were all put on hold while the government moved towards actually reintroducing a complete population register. This plan, which its creators, Peter Shore and the Statistical Policy Committee, saw as meeting "the need for radical and urgent changes," was formulated in a green paper drafted in 1969 and called *People and Numbers*.[30] This paper proposed and explained two different schemes. The first would have been a full-blown population register that recorded basic personal details about everyone in the country. Such a register could exist as a separate data point, or the numbering system it used could be shared across other departmental filing systems to create a linked system, a virtual data matrix, of nominally separate files. The second alternative would have been for the government to use such a common-numbering system but without a population register. This could have delivered many of the advantages of the former system, but would have lacked the central indexing function of the register. This would have meant it would have relied on departments sharing updates on the personal details of members of the public with whom they had contact.[31]

Either version of *People and Numbers* would have swept aside all the political norms surrounding government data and as such would have instigated the data revolution that Wilson referred to in his RSS speech. This was clearly the leading edge of the government's strategy. However, on 8 September 1969, the plan was shelved and the *People and Numbers* green paper was not published due to the wholly contingent arrival of a political campaign to protect personal privacy. It must be emphasised that this retreat was not due to any change of heart within the government, but was rather a reaction to events outside its control. Indeed, when confronted by the privacy campaign, Wilson did not jettison *People and Numbers* but instructed that it should be held in abeyance for a year until after the general election. Labour lost this election and the plan, as such, was never resurrected. This setback meant that though the government's strategic aims remained the same, it would no longer be able to engage in a war of manoeuvre that would have circumvented all existing customary restrictions on data flows by introducing a register or common-numbering system through Parliament. Instead, from this point on, it would find itself engaged in a protracted war of position to overcome the same tendencies in British legal–political beliefs and institutional frameworks that had thwarted and frustrated Acheson, Morris and Titmuss.

2

These factors that acted as drags on the government's ability to enhance its gathering and use of data were first, the institutional codes of practice concerning the confidentiality of data; second, the autonomy of professionals, such as doctors, who held data within ostensibly state systems such as the NHS and third, the decentralised nature of the British government's data collection institutions. Though these had been roundly condemned by Acheson, Morris and Titmuss, and though the incoming Labour government shared the aims of these data trailblazers, these obstacles were well entrenched and removing them, or even tackling them, would require a lot of political effort and was not something that would happen as quickly as its advocates would have liked. This section examines how the government viewed and attempted to deal with the issue of the confidentiality of data and the role of professionals while the following section of this chapter considers the government's data-centralisation drive.

When Peter Townsend's student came up against the National Assistance Board, before the election of the Labour government in 1964, he was attempting to access data from a department that had "insisted throughout its history that, in principle the information it holds about persons who have applied for or enquired about assistance is confidential." Thus, the policy that had been consistently maintained since the department was established in 1934 was that: "information regarding an insured contributor is not supplied to other persons or bodies except with the prior consent of the person concerned or a subpoena."[32] However, the department noted that: "there is no statutory basis for this policy," rather it was "based on general government policy in relation to social-security records, which dates back to 1911."[33] This meant that the guidelines could, at least in theory, be reinterpreted in the light of changes in attitude or policy, but the fact that this could occur did not mean that it would, and neither did it mean that such a change would be easy to obtain. Thus, in 1969, the minister at the Department of Health and Social Security (DHSS) was told that even moves to discuss making improvements in social statistics, which it was agreed were "incomplete" and "lacking in coordination," would involve "issues of confidentiality and privacy," which were described as "awkward."[34]

Such issues were undeniably awkward for ministers and civil servants who wanted change. However, as far as social scientists outside government were concerned, the issue of confidentiality was more serious, in fact it was at the top of a list of the "intractable" problems that prevented their accessing government data.[35] Here the Social Science Research Council lambasted the "existence of sweeping restrictive legislation and – quite often as draconian – the quasi statutory, enforcement of longstanding ministerial faits."[36] All of which was made even more frustrating by the fact that these were not even applied uniformly across Whitehall. Indeed researchers could encounter wide differences between the stringency with which rules were applied, not only between departments, but also between sections of a department with the result that access was often dependent on the whim of the official contacted. There was, in simple terms, an absence of any "coherent policy."[37]

However, once the government's plan to introduce a population register had been withdrawn, it was obliged to pursue its fervent desire for data through this system. Therefore, it also needed to attempt to address the system's practices and reshape its institutions. This brought the government up against this issue of the British legal–political tradition of confidentiality. In fact, this issue had arisen even before *People and Numbers* had been taken off the agenda. In 1967, Claus Moser had been informed that the government saw

> advantages in offering less binding engagements to those who supply statistical information and think that it would be useful to review the pledges of confidentiality that have been given to see whether they need to be maintained in their current form.[38]

Similarly, a few months earlier, the Statistical Policy Committee had concluded that, rather than treating all government data in a similar way, and therefore subjecting it to identical codes of confidentiality, it would be better "to deal with particular problems as they arose."[39] The DHSS also thought that the issue needed to be confronted because "it is clear that problems of confidentiality will persist and new ones will arise."[40] The discussion in this section uses the DHSS's approach to the issue to exemplify the type of analysis and actions produced and pursued by the government since, as the department itself noted in 1972, it collected "a larger quantity and a wider range of information about a larger number of individual citizens than probably any other government department."[41]

The DHSS found the issue so important that, in 1971, it set up a working party to enquire into its confidentiality practices. In April 1972, this group reported to the department's ministers, with its report tellingly stressing that, since the working party had been an informal body whose existence had not been publicised, its report should not be published.[42]

The DHSS had been created by the fusion of two other departments in 1968 and the working party focused its attention on the relationship between the two, previously separate, sets of health and social-security data that had been brought together by this merger. Social workers and others, medical researchers such as Acheson, for example, wanted and needed access to medical records but could not always gain this because doctors held these records protected behind the veil of patient confidentiality.

The Seebohm Committee had reported that it had been repeatedly told about the difficulties this caused and concluded that though "the tradition which leads to the present difficulties is understandable ... it derives from views of professional practice that are increasingly anachronistic."[43] Peter Miller and Nikolas Rose have noted how social government, of the type advocated by Acheson, Morris and Titmuss in the post-war period and introduced by the government in the 1960s, was expert government.[44] These experts, doctors in this case, were needed by biopolitical administration since they alone held, or could analyse, the knowledge that social government needed. However, in order to function as experts, it was necessary that government respect the autonomous professional practices of these holders and manipulators of knowledge.[45] This did not necessarily mean that these experts would be allowed to stand aloof from government rather it meant that while being included within the state, these experts would be able to "generate 'enclosures', relatively bounded locales or fields of judgement within which their authority is concentrated, intensified and rendered difficult to countermand."[46] Thus, experts, such as doctors, found themselves in a position of relative autonomy vis-à-vis the state. But the important point is that the arrival of social government did not conjure these experts into existence, rather the government was obliged to reconfigure and renegotiate existing sets of relationships and enclosures. In other words, it was required to enter into what Michael Moran has called a series of "shotgun marriages" with existing experts, their professional bodies and various state institutions, some of which would have strong attachments to attitudes and professional practices that, being "anachronistic," did not sit harmoniously with the evolving biopolitical ends of social government.[47]

All of this meant that the process would be subject to complexities, contingencies, local variables, illogical twists and setbacks with the government needing to engage in a near constant process of repositioning in its relationship with the medical profession and elements of the civil service.[48] In other words, these were processes that consumed both time and political effort as was witnessed by the length of time it took the government to get hospitals to use standardised forms for their medical records (see Chapter 1). Thus, in 1971, the DHSS working party set up to find a way through the confidentiality morass, reluctantly concluded that it was necessary to avoid being seen "to be imposing standards on the professionals." This was not simply for the sake of political appearances, but because doctors had considerable latitude in deciding the quantity and quality of data they recorded about patients "and if they became of the opinion that their written records were insufficiently protected they would record less – which would be unfortunate both from the operational as well as the research viewpoint."[49] The doctors could protect their professional enclosure by arguing that if confidentiality safeguards were weakened, patients would confide less in their GP, thus threatening the health services that the government was not only politically responsible for, but also paid for.[50] However, the government was also responsible for, and wanted health research. Caught in this cleft stick, it was forced to engage in a political war of position by issuing the medical profession with general confidentiality guidelines, rather than detailed rules, while at the same time chivvying groups and individuals and chipping away at the hitherto strong boundaries of their enclosures.

Thus, in 1971, this working party noted that medical confidentiality did not apply in all cases presented to doctors and clinics. For example, it did not apply when tracing the contacts of someone infected with venereal disease, and this seemed to hold out some hope that general standards of confidentiality could be lowered to this level.[51] However, later the same year, the MRC Sub Committee on Epidemiology reported to the working party that it had reluctantly agreed

> that it must take account of the present climate of medical opinion, particularly as represented by the BMA [the British Medical Association], and recommend that clinical data should only be made available for research purposes to medically qualified research workers.[52]

The subcommittee recognised that there were good reasons why social scientists might need access to such data, and also that social-work records, perhaps just as sensitive as those kept by medics, were not accorded this level of protection, but nevertheless "for the present felt it had to keep to the more limited approach."[53]

However, not all enclosures were so well buttressed against all encroachments. The fifth meeting of the DHSS working party discussed the way social-security offices shared data with the police. There had obviously been some friction around this issue and it was decided that the norm should be the "more permissive practice of the former National Assistance Board."[54] In other words, that frontline staff should stop shielding their clients by claiming confidentiality of their data. The social-security authorities needed the cooperation of the police, in dealing with fraud or handling disturbances in their offices, for example, and it seems that the police expected information in return for this and the authorities were willing to provide it. But what the data modernisers needed was not a series of ad hoc arrangements like this, but rather a device to permit a strategic advance against British data traditionalism.

The DHSS working party's final report indicated the way forward here when it picked up a phrase used in meetings to describe the way in which data provided by an individual could be used across the department, or indeed beyond. This was the concept of "the task." The report concluded that:

> a tacit agreement exists (or may be deemed to exist) between the individual and the person to whom he supplies information that it will be confined to the recipient or passed only to such other persons who may need to know in order to complete the task.[55]

This may sound as though the department was upholding the strictures and traditions of traditional British confidentiality norms; however, a lot hinged on how the word "task" was defined. Here the report admitted that "the boundaries of the concept are unavoidably fuzzy," which meant that decisions regarding the sharing of an individual's personal information across enclosures within the DHSS would be made "where this is considered to be in his interests … in the judgement and on the responsibility of those providing services."[56] In other words, those seeking to weaken existing codes of confidentiality would be judge and jury in their own case.

An indication of how this judgement might be exercised in practice can be gleaned from the immediate reason for this working party being established in the first place. It was set up to address issues of confidentiality arising from a feasibility study of a system of medical-record linkage of precisely the type that Acheson had advocated. This system was designed to replace the existing, "somewhat haphazard and incomplete procedures for linking manual hospital records," with an automated system.[57] This would, it was maintained, "create new and significant opportunities for research" and because of this, it would be of benefit to patients and therefore, by extension could be deemed to be within the boundaries of the task.[58] The committee argued that the benefits to be accrued from this system were so obvious and so vast that they would trump any suggestion that they be referred to "an 'ethical committee' or tribunal" neither of which would therefore be required.[59]

Acheson would, of course, have wholly agreed. Like many arguing for greater access to government data, and for this data to be reformulated, he saw the fundamental ethical question as being how to resolve conflicts between the "long term interest of medical science (and therefore the community)" and "the immediate interest of the patient[s] concerned" in protecting their own confidentiality.[60] This idea, of balance, may give the impression of neutrality between positions, almost of a middle position of stasis. But the reality is that the term "balance," as used in these circumstances, really denoted a characteristically modern restlessness, a system in the unceasing flux of a rebalancing or a repositioning based on the constant arrival of new forces, systems, technologies or knowledge. Indeed, just as it is obviously the case that it takes two separate things to achieve any balance, so the term "balance" was not used in discussions of government data before the 1960s as traditional British notions of confidentiality were so prevalent that there was nothing against which they could be balanced and such views were able, for example, to secure the abolition of the war-time National Register largely unchallenged.

The government, moreover, was well aware that its actions were introducing substantial changes to British politics. Hence when Alec Cairncross (head of the government's economic service 1964–69) wrote to Moser in 1967 about the issue, he recognised that this involved a "change of attitude" and argued that this should be encouraged with more emphasis being "laid on the positive value of the statistics collected and the uses to which they were put and less on the negative pledges to those supplying them."[61] Donnison agreed arguing, in 1972, that: "the 'right to privacy' and the 'right to know' must always be considered together."[62] Whilst Moser argued that there was no need for the government's public presentation of its data plans to draw attention to the issue of confidentiality since any risks could easily be overstated.[63] Clearly, if the balance was going to be swung, the government would be putting all its weight on one side to do so.

The pace and the extent to which the traditions of the post-war British system were challenged can be seen in the DHSS report on confidentiality. Here "the central problem" remained how to determine the correct relationship between individual consent and "the value to the public of the conduct of research." However, by this time, attitudes regarding data had shifted so much that infringements relating to an individual's expectations of confidentiality, by which the report seemed to mean the traditional British view, were now recast by the department as being merely "technical infringements." These, the report maintained, were worthy of note but, it added, "we do not wish to over emphasise them for they are of a kind which constantly arise in matters of government."[64] The point to note here being, of course, that such matters only arise constantly in government processes when the processes acknowledge they exist as a challenge rather than accepting them as being part of the natural terrain of politics, and it is clear that by 1972, the DHSS had shifted its position on these matters. Thus, it had moved from intuitively accepting and almost instinctively working within them, to a more modern understanding, which saw and portrayed them in an increasingly negative light and argued for a system "which makes the maximum use for legitimate purposes of information held within the Department."[65] It thus advocated a system where the definitions of legitimacy, and what constituted the best interests of the public would be decided by the view of the officials concerned rather than by the traditions of Britishness.

3

As has been seen, the ideas of the Wilson government reflected and reinforced those of the data-hungry zeitgeist pioneered by social scientists and medical researchers. This attitude put British governments onto an epistemological, political and administrative track where they would be ever driven to pursue an increasing depth and breadth of population data. The object of this pursuit was a chimera, but Foucault draws attention to how it would impel the state further on the trajectory of becoming "a centred, and centralising power."[66] As has been seen, the government would have to engage with both the enclosures of professional groups and codes of practice that governed the confidentiality of government data. But ministers and officials would also have to engage existing state structures in a war of position in order to build a centralised apparatus that could collect, collate and command the state's data. Moreover, as the data inevitably mounted, so would the centralisation of these institutions. This section analyses this centralisation.

Claus Moser was appointed as the chief statistical officer in 1967 and his appointment is indicative of this centralisation since he was the first holder of that post to also be head of the Government Statistical Service (GSS). But, more importantly, his ideas echoed and emphasised those of all the key members of the Labour government and he was appointed to deliver on these. He was one of the architects of the *People and Numbers* plan, an opponent of the traditional British norms of data confidentiality and saw the "central purpose of government statistics" as being progress in building "integrated systems."[67]

The nature of the organisational structures that would be required to deliver the government's vision for population data was clearly expressed by the Statistical Policy Committee. This body opened its submission on the reorganisation of the GSS by stating that: "there is a need for a much greater degree of centralisation both in the collection of statistics, economic, industrial and social, and in the overall arrangement of the Government's Statistical Services."[68] This was written while *People and Numbers* was still in gestation, and in many senses, the committee's desired level of centralisation could have been achieved had the green paper's plan come to fruition since its central register would have drawn all other datasets into its orbit. However, this plan was not realised. Moser agreed that unitary organisations could offer a great many advantages and in 1978 wrote that: "if we had been starting de novo I would have argued for such a system."[69] However, once *People and Numbers* had been abandoned, Moser had lost the chance to have anything like a clean slate and it would be him, as head of the GSS, who would from this point on have to spearhead the subsequent reorganisation of the government's data machinery and in so doing engage in a series of positional struggles.

The first of these struggles would be to increase central government's power over data gathering. To do so, it would be commensurately necessary to rein in local government. As was seen in Chapter 1, after the war, the National Assistance Board moved to prevent its Liverpool arm from assisting Thomas Simey's students from studying its procedures. But it was only able to do this after the practice had been in place for some years. In other words, the centre could be ignorant of what actually happened in its local offices. Indeed, in 1965, the National Assistance Board could "only guess at the amount and content of the interchange of information that goes on locally," between its offices and the police.[70] Many of the data systems the government relied on, such as the electoral rolls, driving licences and later the poll-tax registers were organised locally. Moser, and others at the centre, seeking to extend the depth and breadth of their resources would argue that this localism should be maintained in order to legitimise these sources of population information. But, as a point of principle, they also agreed with the Seebohm Report in arguing that the centre should always control the precise nature of what was collected and even how it was held. In fact, if this locally collected data were to be collated at the centre, such standardisation was essential. Thus, at a meeting of the CSO's Committee for Statistics for Social Policy (another recent addition to the data firmament, established in 1967 by Moser who would be its chair), held to discuss local authority social services statistics, Moser argued that any mention in DHSS guidance to its local offices that they could compile locally based registers of their clients should be removed "as soon as possible while the system for personal social services statistics were still being developed."[71] Clearly, when the ink dried on this guidance, the message it delivered would be one of increased centralisation.

However, the institutional struggles waged by Moser and others at the centre were not confined to nipping local initiatives in the bud, since the centralisation drive would be made all the harder by the nature of the GSS itself. At a staff conference in 1976, a member of the GSS presented a paper called "thinking aloud," about the GSS budget that opened by stating that in order to consider the organisation's budget, it was necessary to define the organisation itself and "this is not as simple as it sounds."[72] Moser at the GSS's apex agreed noting, two years later, that the GSS was "a dispersed and somewhat splintered organisation."[73] This was because the service, that the government was trying to repurpose to meet its new biopolitical agenda, was a product of very different political traditions and consequently its institutional forms were ill suited to the new demands placed upon it.

The government's Statistical Policy Committee was both a product of these centralisation processes and, once it was at work, sought to accelerate these tendencies. The committee was established in April 1967, and though its first chair was Peter Shore, Wilson took the chair at its first meeting "to show that I meant business."[74] Within a matter of a few months of starting work, Shore told Wilson that: "there is now widespread recognition by departments of the need for greater centralisation of statistical work and for strengthening the role of the Central Statistical Office."[75] This was, he further argued, because: "there can be little doubt that the present shape of the Statistical Service and the distribution of resources within it reflect historical circumstances rather than current needs."[76] What this meant in practice was that GSS statisticians were dispersed across Whitehall, embedded in a wide variety of departments, where they served the needs of their host ministries as and when required. The centre of the GSS, the chief statistical officer in the CSO, was part of the Cabinet Office. The CSO retained responsibility for the professional standards of members of this splintered body, but clearly could not direct their work to match the needs of any centrally determined policy. This, in turn, limited the centre's ability to form or deliver such policies. The system had thus evolved to serve the needs of departments and certainly had not, according to the Statistical Policy Committee, "been consciously planned as a whole to meet the changed needs of modern government."[77] If it had been planned to serve the centrally determined needs of modern government, the main obstacle produced by this system, identified by the CSO as "departmental safeguards which … impede the effective statistical use of the data in other parts of government," would certainly not have been built into the system.[78]

This issue, of confidentiality, which prevented the sharing of data across departments was, as has been seen, a serious impediment for those who wanted to see a step change in government data. But unfortunately for them, it was not the only problem that came with the existing government data system. A 1968 meeting of the CSO's Committee on Statistics for Social Policy listed these issues under three headings. The first was that the system encouraged a lack of communication between departments that were responsible for processing their own data for their own purposes (the "data islands" referred to later by Tony Blair). This, it was argued, led to the situation where data might be duplicated across different departments, or, where one department might hold data that it did not publish leaving others in ignorance of its existence despite the fact that they could have benefitted from it. Congruent on this haphazard distribution of information, there was also likely to be an iniquitous distribution of statistical resources, with some departments being oversupplied with statisticians. The second problem derived from the existing system was a low usage rate of information. A department might hold a stock of population data but only extract from it the information it needed for its own purposes, leaving the rest of this stockpile unused. The third problem inherent in the system was that different departments used different classification systems meaning that data streams running through Whitehall could not merge.[79]

As an example of these problems, the committee discussed the National Food Survey (NFS) that enquired into the nutritional standards of the food consumed by ordinary people. This had begun during the war when all concerned agreed that this data was extremely important, however, no one could come up with a reasonable justification for why these surveys were still being conducted in the late 1960s. For one thing, it was widely accepted that no one in Britain was at risk of being malnourished, for another there were, by this time, other surveys such as the Family Expenditure Surveys collecting the same type of information. Moreover, it was later conceded that since the NFS had a non-response rate of 45%, its findings were not even representative of the population as a whole.[80] In the final analysis, this appeared to the committee "to be a classic case of resources continuing to be devoted to an enquiry whose main results no one really wants, in the last resort because it has been no one's responsibility to suggest a reallocation."[81] In other words, the interrelated problems of the British government's population data existed in a reciprocal relationship with the institutions that held this information. In this sense, then, the root of the problem was seen to be an absence of centralised oversight and control.

Therefore, for Wilson, Moser, Acheson and Titmuss, and all other advocates of accelerating the government's data traffic along Whitehall, the obvious solution to these bottlenecks, dead ends and snarl ups was centralisation. In the late 1960s, the key project in this regard was *People and Numbers* the common-numbering system of which would, it was suggested, overcome most of the problems of the decentralised patchwork system operational at that time. But the *People and Numbers* plan was abandoned and, by 1973, Moser had come to terms with the fact that he would have to work with the structures he had inherited, rather the ones he would have designed de novo. Once in this position, though he continued to argue that the balance of forces within the GSS was not exactly how he would have liked, he maintained that its decentralised structures could nevertheless offer advantages.[82] The main of these was that statisticians were at the heart of policy-making in departments, they were not, as Wilson termed it, "back room boys."[83] As Moser put it, GSS statisticians were not confined to the production of pure facts, they were "part and parcel of the interpretive and decision-making arm of government" and located here, in policy-making teams, they were in a prime position to ensure that the information flow was meeting the requirements of policy and to provide analysis of data to impact policy-making decisions.[84]

In March 1974, less than a month after Wilson, returned to Downing Street, Moser wrote to update him on the GSS's work. He was pleased to note that the service had boosted its publication of data with *Social Trends* joining five other new series covering the fields of labour, taxation, health and social-security statistics, in becoming annual publications; that it had launched the General Household Survey in 1970; was planning a scheme of medical-record linkage and was developing a national cohort study. He also told Wilson that though he was working with the decentralised system and to some extent at least could see advantages in doing so, he was nevertheless pursuing a strategy of centralisation with the CSO assuming increased managerial control over the whole GSS. Part of this had been realised through a 1972 reorganisation of the CSO itself into five divisions: two concentrating on economic statistics, one on social statistics, one on computerisation and publications and one running in-house management.[85]

However, though these changes may have increased the public profile of the GSS and helped to increase the efficiency of its centre, the rest of the organisation still remained dispersed, power over data gathering still remained scattered across Whitehall and there were still institutional positions to be contended. One of these concerned authority over the conduct of surveys. In 1972, the government received the report of the Committee on Small Firms (the Bolton Committee) that gave a chapter of its report to the subject of the burden placed on small firms by government form filling. The report recommended that the Survey Control Unit of the CSO should be given the power to veto or amend any survey planned by any department. Neither Moser nor the prime minister, Edward Heath (in office from June

1970 to February 1974) thought that a formal veto was an appropriate level of authority for the CSO. However, Heath wrote to all ministers in charge of departments telling them that he "attached great importance" to their consulting the CSO whenever they were planning to conduct a new survey, or to significantly alter an existing one. Adding that such consultation should always occur at an early stage in the development work, a timescale that would, of course, have maximised the CSO's ability to intervene and given it a de-facto veto over such plans.[86] This, obviously, illuminates one of the ways in which the power to gather data was dispersed across Whitehall, but the difficulties experienced by Moser and others seeking to centralise power over data within this system is made plain in this case by the fact that four years later, a different prime minister, Jim Callaghan (in office from October 1974 to May 1979) had to send a virtually identical instruction to his ministers.[87] Indeed, this struggle to control data collection did not even stop at that point (1976). Thus, when Moser retired in 1978, his final report on the condition of the GSS noted that its staff "should have the opportunity to participate in the construction of administrative forms and data sources," which implied that they were still not, after six years of trying, actually doing so.[88]

Once *People and Numbers* had been abandoned, Moser was thrown back on other elements of his reform programme, which in the absence of the overarching scheme of a population register now assumed greater importance. This included reshaping the institutions responsible for gathering and circulating government data. The most important new arrival in Whitehall in this period was the Office for Population Censuses and Surveys (OPCS) and, by November 1969, Moser was extremely anxious to get this body into action at the earliest possible opportunity. He envisaged the OPCS as a crucial link in the centralisation process "the point of the office," he argued, was "to integrate census and surveys work as far as possible," adding tellingly that: "all departments in Whitehall will, I imagine, be in its climate."[89] The office was eventually set up on 11 May 1970 by a merger of the General Register Office and the Government Social Survey Department. Wilson had mentioned this in a speech in 1969 and, when he formally announced this merger in Parliament, he stated that this was being undertaken to improve the quality of the social statistics that were vital to the formulation of policy. Moreover, he added that though the OPCS would be in the portfolio of the secretary of state for social services and run by the registrar general, he attached such "particular importance to this development" that he would personally "assume oversight of the general policy of the new office."[90] The OPCS was, Wilson told Parliament, to be "closely linked" with the CSO and the heads of the two offices were to work together under ministerial guidance to direct surveys and statistical work. The Heyworth Report (see above) had argued that the social survey needed to be given more institutional/political heft in its dealings with other Whitehall offices and here Wilson delivered on that recommendation.

Background briefing notes made plain why Moser was so keen to have this new office engaged in its work. These stated that he would chair the policy committee, the link between the OPCS and Moser's base in the CSO, and that he would have responsibility for presenting new programmes of work. Thus the creation of the OPCS, it was made clear, was part of the process of "strengthening the influence of the CSO on planning and programing."[91] In other words, it was conceived as being part of the process of developing the "centred, and centralising power" inherent in Britain's data turn.[92]

Looking at the work of the OPCS, in 1987, three scholars remarked that: "much of its work resembles that of the American Bureau of the Census, monitoring the population in the sense of providing detailed facts."[93] However, 10 years earlier, Moser had noted that though what he had hoped to achieve was "something that could compare in excellence with, e.g., the Bureau of the Census in Washington," this desired outcome had not materialised.[94] This discrepancy between aim and reality highlights the strengths of expert enclosures and the difficulties faced by those like Moser who, being unable to begin their plans de novo, were obliged to work with the materials to hand. These difficulties in this shotgun marriage came to a head in 1976 when Philip Redfern, the deputy director of the OPCS, conducted an enquiry into "the future of the office." Redfern would later write that the problems at the OPCS stemmed from two sources. The first of these was the unhappy union between the pre-existing bodies, the GRO and the government social survey, that had been brought together to form the OPCS. Second, there were the "external strains too in relations with the CSO arising from the CSO's perhaps understandable wish to have a (if not the) key role in making statistical policy." Here, Redfern noted that "to a degree such a strain is inevitable in a decentralized statistical system like ours, in which the statistician looks for guidance in two directions: to his political master and to the Head of the GSS."[95]

Inside the OPCS problems quickly arose, Redfern argued, due to the very different political cultures of the two parties to the merger that had formed the new office. One of these, the GRO "had not changed since it first began in 1837."[96] Indeed, despite its name carrying echoes of continental bodies that executed biopolitical population registration, Edward Higgs has shown how the GRO was established to carry out its work in very different ways to its continental counterparts because it existed to provide data for different reasons. The GRO focused on providing information that would guarantee the inheritance rights of legitimate heirs.[97] It was, in other words, embedded in British political traditions. However, "the government social surveys [which] had developed rapidly since they began in WW2" followed a path much more conducive to Moser's worldview.[98] These fundamental political differences, Redfern argued, meant that the merger between the two had not worked and the OPCS had not developed a clear unified purpose. This was exacerbated by three other factors. The first of these was that each of these components also had different office traditions and working practices that led to strains and conflicts between them. Second, the components of the GRO had long-established connections with a variety of government departments (the Medical Statistics Division with the Department of Health, for example) while the social survey, operating across Whitehall, did not. Third, the merger had attempted to bring together bodies that were geographically dispersed across the country. The headquarters was in London, the executive arm in Titchfield and the NHS Central Register in Southport. The net result of this was, Redfern maintained, that as a unit, the OPCS was in reality eight divisions, each of which pursued a more or less independent policy. Thus, it did not really constitute a coherent organisation.

Redfern's report argued that the solution to this was to give the OPCS more meaningful work to do. In his view, the disparate nature of the functions performed within the OPCS was made worse by the fact that much of this work focused only on collecting data that "tends to become mechanical with insufficient understanding of the validity and meaningfulness of the results." The OPCS could be given a clearer focus if this were "complemented and enhanced by more work here interpreting the results and comparing them with information from other sources."[99] Redfern's plan would have seen the OPCS develop its functions as a centre for the production, compilation and analysis of social data. He argued that the organisation was "uniquely placed, through its censuses and surveys to contribute to the statistical analysis of inter-relationships between topics … and we believe the OPCS should do more of this analysis, research and interpretation than it has hitherto."[100]

Redfern was not a devotee of the data traditions of British political culture. In fact, as will be seen, he was an ardent, if not indeed a strident, advocate of the population registration policies forwarded by Moser. Indeed, it was precisely because he saw population data in decidedly modern, positivist terms, that he maintained that its interpretation was best left to a body of experts, such as the OPCS, which could not be seen as the plaything of politicians.[101] The OPCS was run by the registrar general, and Redfern argued that it ought to be renamed "The Office of the Registrar General."[102] This was more than a piece of cosmetic rebranding and Redfern chose this title because the registrar had formal independence from ministers and government departments, and this, he argued, reassured the public that the privacy and confidentiality of information collected by the registrar (such as the census) would be protected from political misuse. Indeed, he added, that doctors had argued that this reassurance, as provided to cover medical records, could not be maintained if the registrar "were to be too closely identified with, or become a part of a government department, such as the CSO, that was directly responsible to a minister."[103] Redfern's report thus forwarded doctor–patient confidentiality to protect the expert enclosure of the OPCS from what he saw as the overzealous attention of the CSO.

Moser had had a clear aim in creating the OPCS; he had wanted to centralise and streamline population data. Instead, his inability to initiate plans de novo and his reliance on a shotgun wedding that united existing institutions produced another enclosure from within which experts, such as Redfern could, while ultimately arguing for his population data data policies, actually thwart the centralisation that Moser saw as necessary to their implementation. Moser, in his own words took three months to "brood" over the OPCS report and when he replied to the registrar general (George Paine, he held this post from 1974 to 1978), who had not produced the paper, but was, nevertheless, responsible for Redfern and the OPCS, he assured him that his objections were not personal and would not impinge on their long-standing friendship, they were however intensely political.[104] These objections can be placed in two groups.

The first of these concerned the report's defence of the decentralisation of British data institutions. In 1978, looking back on the way the GSS had changed during his time in office, Moser reflected that he had been given "far greater difficulties in achieving a full co-ordination of statistics and statistical activities than I envisaged" because of the "semi-centralised" nature of the system he had initiated.[105] This could well have been partly a reference to this relationship with the OPCS. Here, Moser saw Redfern's report as an attempt to create an independent OPCS and while not suggesting that his own scheme was beyond improvement, he was "sure – and all my colleagues in other countries would agree," that to be called an improvement, any change to his plans "would have to be in the direction of more centralisation, not less."[106] This centralisation was, Moser argued, not a power grab by the CSO, rather it was a reflection of underlying trends in modern society where economies, societies and governments were becoming increasingly interlinked "and the less integrated and co-ordinated the statistics are, the less use they will be in depicting conditions and trends."[107] Moser had intended the OPCS to be a cornerstone of the system he had devised. He maintained that he had never sought, and did not want to assume power over the registration functions within the OPCS, but insisted that since the CSO was one of the OPCS's main customers, in that it received and deployed OPCS data productions, it had a right to be represented within the OPCS and to determine the nature of its output. The OPCS report had presented the CSO's involvement in its work as being heavy-handed and unnecessarily bureaucratic. But Moser did not accept this as a fair representation of the way he managed the steering committee that navigated between the CSO and the OPCS. However, he did agree with Redfern that inevitably any "attempt to produce some central management in a decentralised system produced tension." He even added that he saw "no harm in this," but then forwarded a caveat that this could only be the case if "the parties respect the validity of each other's roles, and each other's skills and energy."[108]

However, Moser's second bone of contention with the OPCS report precisely concerned the skill and energy with which the OPCS discharged its work; here he made two points. First, he agreed with the report that the OPCS remained a disparate organisation, a collection of parts, rather than the sum total of them. But he argued that this was because of a failure of the OPCS's management to facilitate the merger of its constituent units. Second, he noted bluntly that the management of the CSO had "lost confidence in the skills and energy of the management of the OPCS because of failures in important areas of statistics."[109] Here he presented a catalogue of errors that included: a delay in the output of the 1971 census, poor population forecasts and discrepancies in Scottish migration figures. Thus, he concluded, if relations between the CSO and the OPCS had deteriorated, it was because he, as head of the GSS, bore the brunt of criticism for these shortcomings. In short, were the OPCS to achieve the independence the report sought, it would be a retrograde step for Britain's population data.

That Moser should view the OPCS report with such alarm shows how important the data turn of British politics was to him and by extension to the government that had appointed him. He had been "most disturbed" by the loss of the *People and Numbers* plan but, once this opportunity had passed, he dug in to engage in a war of position to push the government's data systems forward against codes of confidentiality, professional enclosures and decentralised structures.[110] The policies, systems and institutions he and the government used to do this, and to broaden and deepen government population data, are examined in the next chapter.

Notes

1 TNA, BN 102/1, To Mr Windsor, *Disclosure of Information*, n.d., ca. 1965, 1 and 7.
2 TNA, BN 102/1, National Assistance Board, *Memorandum 1356: Disclosure of Information for Research Purposes*, 17 Mar. 1966, 1 and 2.
3 TNA, CAB 139/742, *Draft Speech for the Royal Society of Statisticians Banquet on the Occasion of the 37th Session of the International Statistical Institute*, September 10, 1969, 11.
4 Harold Wilson, "Statistics and Decision-Making in Government – Bradshaw Revisited," *Journal of the Royal Statistical Society. Series A* 136 (1973): 4.
5 TNA, BN 29/1820, Home Office Research Advisory Committee, *Cohort Studies, Notes by Dr. W.D. Wall*, n.d., ca. Dec. 1965, 2.
6 TNA, PREM 13/1432, T. B., *Economic and Social Intelligence*, 30 Nov. 1964, 3.
7 E. D. Acheson, *Medical Record Linkage* (London: OUP and the Nuffield Provincial Hospitals Trust, 1967), 135, 183, 2 and 52.
8 *Report of the Committee on Social Studies (Heyworth Committee)*, Cmnd. 2660 (June 1965), 43.
9 Ibid., 43–44.
10 *Report of the Committee on Local Authority and Allied Personal Services (The Seebohm Report)*, Cmnd. 3703 (July 1968), 144.
11 TNA, CAB 139/742, *Extracts from Speech by the Prime Minister the Rt Hon. Harold Wilson, O.B.E., M.P. at the International Statistical Institute Banquet, Wednesday 10th September, 1969*, 4; and, TNA, CAB 139/751, *Consideration of Social Statistics by the Ministerial Committee on Social Services*, 20 Jan. 1970, 1.
12 TNA, CAB 139/567, Statistical Policy Committee, *Reorganisation of the Government's Statistical Service, Interim Report by the Chairman of the Statistical Policy Committee*, 11 Aug. 1967, 1.

13 TNA, BN, 89/202, CSO, Committee on Statistics for Social Policy, *Third Meeting*, 9 Nov. 1972, 6.
14 TNA, CAB 164/1695, IT Advisory Panel, 1 Aug. 1983, 1; Philip Redfern, "A Population Register or Identity Cards for 1992?" *Public Administration* 68 (1990): 510, and Philip Redfern, "Source of Population Statistics: An International Perspective," *Population Projections: Trends, Methods and Uses* (London: OPCS Occasional Paper, 38, 1990), 112.
15 The Cabinet Office, *Transformational Government Enabled by Technology*, Cm. 6683 (Nov. 2005), 5.
16 Philip Redfern, "Precise Identification through a Number Protects Privacy," *International Journal of Law and Information Technology* 1 (1994): 308.
17 Mr. F. E. Whitehead, in, Philip Redfern, "Population Registers: Some Administrative and Statistical Pros and Cons," *Journal of the Royal Statistical Society, Series A* 152 (1989): 30.
18 Michel Foucault, *Security, Territory, Population* (New York: Picador, 2004), 78.
19 Ibid., 78 and 79.
20 Michel Foucault, *Society Must Be Defended* (London: Penguin, 2004), 62.
21 Ibid., 61, and Whitehead, in, Redfern, "Population Registers", 30.
22 *The Seebohm Report*, 146.
23 Wilson, "Statistics and Decision-Making in Government", 7.
24 Ibid., 144.
25 TNA, BN 102/1, to Herbison, 31 May 1966, 5 and 2.
26 TNA, PRO 30/87/61, David Donnison, *The Contribution of Research to Social Policies and Programmes*, Feb. 1972, 15.
27 *The Seebohm Report*, 145.
28 Acheson, *Medical Record Linkage*, 1.
29 TNA, RG 28/306, *Note from Mr. West re: Proposals Involving Staff Increase in Registration Division*, 16 May 1966, 1.
30 TNA, CAB 134/3274, Statistical Policy Committee, *Meeting 13 July 1967*, 4. For further information on *People and Numbers*, see Kevin Manton, *Population Registers and Privacy in Britain, 1936–1984* (London: Palgrave MacMillan, 2019).
31 TNA, CAB 130/427, *Draft Green Paper: People and Numbers*, 12 June 1969, 5.
32 TNA, BN 102/1, *Disclosure of Information*, n.d., ca. 1965, 1 and 2.
33 Ibid., 1.
34 TNA, CAB 139/561, *Brief to Minister of State, Department of Health and Social Security, 'Consideration of Social Statistics'*, 16 Dec. 1969.
35 TNA, BN 102/3, Social Science Research Council, *Newsletter*, June 1970, 1.
36 Ibid.
37 Ibid., 3.
38 TNA, CAB 139/566, Cairncross to Moser 2 Nov. 1967, 1.
39 TNA, CAB 134/3274, Statistical Policy Committee, 13 July 1967, 3.
40 TNA, BN 102/8, to Secretary of State, *Problems of Confidentiality and Personal Records*, Apr. 1972, 1.
41 TNA, BN 102/4, DHSS, *Confidentiality of Personal Records, Report of Working Party*, Apr. 1972, 4.
42 TNA, BN 102/8, to Secretary of State, *Problems of Confidentiality*, 1.
43 *The Seebohm Report*, 199.
44 Peter Miller and Nikolas Rose, *Governing the Present* (Cambridge: Polity, 2008), 108.
45 Ibid., 203.
46 Ibid., 209.
47 Michael Moran, *The British Regulatory State: High Modernism and Hyper-Innovation* (Oxford: OUP, 2003), 179.

48 Nikolas Rose, *Powers of Freedom: Reframing Political Thought* (Cambridge: Cambridge University Press, 1999), 210.
49 TNA, BN 102/5/2, Working Party on Confidentiality of Personal Records, *Minutes of the First Meeting*, 29 Sept. 1971, 4.
50 TNA, BN 102/5/2, Working Party on Confidentiality of Personal Records, *Minutes of the Third Meeting*, 16 Nov. 1971, 3.
51 TNA, BN 102/6, Working Party on Confidentiality of Personal Records, *Common Practices, Anomalies. Problems and Tentative Recommendations*, n.d., ca. late 1971, 3.
52 TNA, BN 102/5/2, Working Party on Confidentiality of Personal Records, 16 Nov. 1971, 3.
53 Ibid.
54 TNA, BN 102/5/3, Working Party on Confidentiality of Personal Records, *Minutes of the Fifth Meeting*, 13 Dec. 1971, 1.
55 TNA, BN 102/4, DHSS, *Confidentiality of Personal Records*, 12.
56 Ibid., 14.
57 TNA, BN 102/5/3, Working Party on Confidentiality of Personal Records, 13 Dec. 1971, 4.
58 Ibid.
59 Ibid., 5.
60 Acheson, *Medical Record Linkage*, 100–101.
61 TNA, CAB 139/566, Cairncross to Moser 2 Nov. 1967, 2.
62 TNA, PRO 30/87/61, Donnison, *The Contribution of Research*, 20.
63 TNA, CAB 139/563, Moser to Reed, 16 Sept. 1968, 2, and Committee on Statistics for Social Policy, Meeting, 9 Nov. 1972, 5.
64 TNA, BN 102/4, DHSS, *Confidentiality of Personal Records, Report of the Working Party*, Apr. 1972, 22–26 and 27.
65 Ibid., 20: and TNA, BN 102/5/3, Working Party on Confidentiality of Personal Records, *Minutes of the Fourth Meeting*, 29 Nov. 1971, 2.
66 Foucault, *Society Must Be Defended*, 61; and, Whitehead in, Redfern, "Population Registers," 152 (1989): 30.
67 TNA, PREM 16/1884, Moser, *Review of the Government Statistical Service*, 4 Aug. 1978, 8.
68 TNA, CAB 139/567, Statistical Policy Committee, *Reorganisation of the GSS*, 1.
69 TNA, PREM 16/1884, Moser, *Review of the GSS*, 22.
70 TNA, BN 102/1, *Disclosure of Information*, n.d., ca. 1965, 10.
71 TNA, BN 89/202, Committee on Statistics for Social Policy, *9 Nov. 1972*, 6. For the committee itself, see TNA, CAB 139/561, CSO: Committee on Statistics for Social Policy, *Terms of Reference*, 25 Mar. 1968, 1.
72 TNA, BA 17/1085, GSS Senior Staff Conference 26-27 Nov. 1976, L. S. Berman, *A, GSS Budget and Questions of Priorities: Thinking Aloud*, 1.
73 TNA, PREM 16/1884, Moser, *Review of the GSS*, 31.
74 TNA, PREM 13/1432, T. B., *Economic and Social Intelligence*, 4 and 5; Harold Wilson, *The Governance of Britain* (London: Sphere, 1977), 90; Wilson, "Statistics and Decision-Making in Government, 6–7.
75 TNA, CAB 139/566, to Prime Minister, n.d., ca. Aug./Sept. 1967, 1.
76 TNA, CAB 139/567, Statistical Policy Committee, *Reorganisation of the GSS*, 1.
77 Ibid., 1.
78 TNA, RG 19/758, Annex: *Confidentiality Guidelines*, Feb. 1973, 8.
79 TNA, CAB 139/561, H. E. Bishop to A. N. J. Baines, 17 June 1968, 18.
80 TNA, CAB 108/769, CSO Committee for Statistics for Social Policy, Working Party on the General Household Survey, *Validation of Statistical Output from the General Household Survey*, 14 Jan. 1971, 3.

81 TNA, CAB 139/561, Bishop to Baines, 19.
82 Claus A. Moser, "Staffing in the Government Statistical Service," *Journal of the Royal Statistical Society* 136 (1973): 75–88.
83 TNA, CAB 139/742, *Extracts from a Speech by the Prime Minister*, 2.
84 Claus A. Moser and I. B. Beesley, "United Kingdom Official Statistics and the European Communities," *Journal of the Royal Statistical Society*136 (1973): 541; and Moser, "Staffing in the GSS," 1973, 76.
85 TNA, PREM 16/1884, Moser, *Some Recent Developments and Future Priorities in the Government Statistical Service*, 21 Mar. 1974, 1–4.
86 TNA, FD 9/1721, *Prime Minister's Minute to all Members of the Cabinet and Ministers in Charge of Departments*, 1 Mar. 1972, 1–2.
87 TNA, FD 9/1721, Prime Minister, *to Ministers in Charge of Departments*, 8 Nov. 1976, 1.
88 TNA, PREM 16/1884, Moser, *Review of the GSS*, 19.
89 TNA, BA 17/1085, Moser to Osmond, 17 Nov. 1969, 1.
90 TNA, BA 17/1085, Office of Population Censuses and Surveys, *Draft Parliamentary Question*, n.d., ca. Nov. 1969, 2.
91 Ibid., 3.
92 Foucault, *Society Must Be Defended*, 61.
93 Aubrey McKennell, John Bynner and Martin Bulmer, "The Links Between Policy, Survey Research and Academic Social Science: America and Britain Compared," in *Social Science Research and Government: Comparative Essays on Britain and the United States*, ed. Martin Bulmer (Cambridge: CUP, 1987), 249.
94 TNA, BA 17/1085, *Comments by Sir Claus Moser, OPCS: Final Report on the Committee on the Future of the Office*, 24 Jan. 1977, 5.
95 Philip Redfern, "Obituary: George (Toby) Paine", *Journal of the Royal Statistical Society* 156 (1993): 121.
96 Ibid.
97 Edward Higgs, *Life, Death and Statistics: Civil Registration, Censuses and the Work of the General Record Office, 1836–1952* (Hatfield: Local Population Studies, 2004).
98 Redfern, "Obituary: George Paine," 121.
99 TNA, BA 17/1085, OPCS, *Final Report of the Committee on the Future of the Office*, Oct. 1976, 4.
100 Ibid., 5.
101 Ibid., 7, 9.
102 Ibid., 30.
103 Ibid., 7.
104 TNA, BA 17/1085, Moser to Paine, 24 Jan 1977, 1.
105 TNA, PREM 16/1884, Moser, *Review of the GSS*, 32.
106 TNA, BA 17/1085, *Comments by Sir Claus Moser, OPCS*, 6.
107 Ibid.
108 Ibid., 3.
109 Ibid., 4.
110 TNA, PREM 13/3257, Moser to Trend, 12 Dec. 1969, 1.

3 Government Data Systems 1964–79

The previous chapter showed how Wilson's Labour Party came into government with attitudes that had previously been the preserve of social scientists and medical researchers, which led them to try to increase and systematise government's holdings of population data. However, though the Labour government and its allies may have wanted to modernise data flows, achieving this was not a straightforward process: after all, attitudes are neither policies nor institutions. In 1973, the Central Statistical Office (CSO) stated that some of the data it would have liked to be able to gather could not be collected by anything "short of a population register."[1] But after Wilson was obliged, in 1969, to abandon his attempt to introduce such a population register, turning these attitudes into reality became even more complex as the government had to engage in a war of position with the existing data system. In doing this, government used the two data strategies brought forward by social scientists and medical researchers; it conducted research to gather information and it also linked the datasets that it already held to form a wider and deeper data matrix. This chapter outlines and analyses how government attempted to use these methods to improve its population data. To do this, it is divided into four sections. The first two each look at a government research project.

The first section considers the 1970 national cohort study. This illustrates how the study was set up to further biopolitical ends by establishing a benchmark standard of what constituted normal behaviour. The section also establishes the lengths to which its founders contemplated going to get this data. The second section looks at the General Household Survey (GHS) that was initiated just after the 1970 cohort study. This research project was directed by the Office of Population Censuses and Surveys (OPCS) and held an annual survey of all the adults in 15,000 households to glean details about their lifestyles. This section details the extent of the GHS's popularity across Whitehall and beyond, and how, like the cohort study, it was used to hone research methods and wear down confidentiality norms.

The third section is the first of the two that explores how the government aimed to increase the use of data it already held. This focuses on the longitudinal study that began in 1971. It explains how this project sought

DOI: 10.4324/9781003252504-4

to know the people by uniting data across the boundaries of government departments through a common numbering system of the type pioneered by Donald Acheson. This discussion also illustrates how the study was used to both sharpen data-gathering processes and to legitimise subsequent data trawls. The fourth section looks at the ways government attempted to link its existing data streams through two wide-ranging projects. The first of these was the Joint Approach to Social Policy (JASP) launched in August 1975. This demonstrates how the JASP was dependent on developing and deploying a data matrix in order to find what government labelled pockets of deprivation and thus to make what Wilson called "an important contribution" to building a "compassionate" society.[2] The political shortcomings of the JASP were responsible for its demise within two years of its launch. However, at the same time, the CSO had a parallel plan to increase the use of data held for administrative purposes by linking these datasets through a common numbering system. This scheme is also examined here.

1

In the previous chapter, it was noted that the government concurred with David Donnison in seeing research as a "national resource." He argued that any government that did not "bring educated minds to bear on its own problems is wasting public money."[3] He saw that researchers could make a lasting and deep impact in government by identifying the themes, issues and problems that recurred across different branches of a department's work. In fulfilling this role, he argued, researchers needed to "walk a tightrope in [their] relations with power." Researchers needed to be close enough to policy makers to be able to contribute to the debate, but needed a strong enough enclosure to ensure that they could defend their own approach and, most importantly, while respecting the questions posed by policy makers, also formulate and pursue their own.[4] The OPCS was established in 1970 and, under Philip Redfern, the office adopted this approach to research and set up a variety of long-term research projects that extracted data by sampling the population. This section considers one of these, a cohort study that began in 1970 and which (in 2021) is still running.

This cohort study was begun as the result of a series of meetings held in 1965. At one of these Dr W. H. Hammond defined a cohort study as: "a longitudinal follow up of a population sample, usually a birth group."[5] Ideally, he added, this would be based on a constant stream of data about the individuals monitored. However, since this was impossible, what it usually involved was a sweep of the cohort at predetermined intervals to unearth information about events that had occurred since the last such trawl. Cohort studies had been pioneered in the United States in the interwar period, and by the 1960s, Britain already had two that were running. The first of these had begun in 1946 while the other dated from 1958. However, by the mid-1960s, as the demand for population data grew, both of these were held to be deficient as they were seen to lack the thoroughness required to meet the demands for population knowledge. The first had been established during the period of post-war austerity and so was a comparatively small-scale survey (it examined 5,000 people from a birth cohort of 17,0000) that was not able to conduct "detailed follow-through of individuals with strictly comparable instruments."[6] The second, the 1958 study, though "infinitely richer in the output of essential information," did not study points such as "parental care, aspects of personal and social development, physical growth and the home environment."[7] It was argued that the new study would fill in the blanks left by these previous efforts. It would provide information on groups of people, "believed to be particularly vulnerable on medical, psychological or social grounds," who would be juxtaposed to a subsample of "normal" children. Were this study to be set up and maintained at this level of enquiry for long enough it would, this paper argued, yield information about "precipitating causes, and about the dynamics of the interaction between personal and environmental characteristics, as well as having reliable population statistics against which to compare individuals and groups."[8]

Other methodologies used to study populations were flawed, it was argued, as they began with a group predetermined to be worthy of study because they were seen as being, in some way, problematic. This meant that they were not, by definition, a normal group and that since the problem they presented (juvenile delinquency, for example), was already known, any attempt to retrospectively trace the causes of this would inevitably be so prejudiced by the knowledge of the way their lives had unfolded as to be meaningless. However, a proper cohort study followed people through their lives and so gathered information before any abnormal outcome occurred. It therefore facilitated the drawing of meaningful comparisons across the whole group, and so allowed causation to be calculated. In this way, the cohort study would see people "in the round and would provide an opportunity for widely based studies involving very different disciplines."[9] To do this, the study was to be run by a permanent interdisciplinary team that would augment their skill set by bringing in specialists as and when required.[10]

This vastly superior scientific research methodology, and "the harvest of knowledge" that it offered, was held to be necessary because society was undergoing rapid changes across a wide gamut of societal fields that interacted with each other.[11] Though some of these were simply products of "more or less uncontrolled processes," some, such as the introduction of comprehensive education, had been produced by the actions of government. It was thus incumbent on government to study how these measures impacted the people and a proper cohort study, which did not focus on looking for the explanation of a known outcome but rather gathered data on the totality of life as it played out, was the only way to achieve this. Therefore, a study designed like this would achieve what Foucault described as an "an endless extraction of knowledge."[12] Moreover, as one of the founding presentations of the 1970 study stated, it was expressly designed to do this in order to "give us at least the chance of acting in time to prevent possible ill-effects."[13] It was, in Foucault's words, "a technique of normalization," operating as part of a systematic "power of regulation."[14] As the National Bureau for Co-operation in Child Care put it, in 1966, "only when there is a better understanding of normal child development and normal learning processes will the normative bases be available for studying and interpreting deviant and maladaptive behaviour in children and young people."[15] This desire to locate the benchmarks of normality was not something initiated by the 1970 study; rather, this study was developed to provide information that had been sought for some time. For example, in 1952, the Social Medicine Research Unit (the renamed SMU, see Chapter 1) reported that one of the main focuses of its work had always been "studying the 'normal' as well as the 'sick' ... because often there are no facts at all about the frequency and distribution of the most important – and elementary phenomena." All of which was necessary, it stated, to expel the "hot air" that usually shrouded any understanding of the essential objects of study for social medicine, such as the family and which left researchers "vastly ignorant" of the realities of life.[16]

As well as producing superior data, the cohort study also held out the promise of refining the methods used to gather such data. Indeed, one of the reports that instigated the 1970 study stated that: "the greatest need, therefore, which a future cohort study could fulfil is to provide a 'laboratory' for the standardising of specialised observational techniques."[17] These statistical and psychometric methods already existed, and it was recognised that they could provide the desired results, but they had not yet been brought together to measure development over time. In this sense then, the cohort study was designed not only to gather population data, but rather to do so in ways that would produce the institutions, techniques and systems that would encourage more such data gathering.

This cohort study thus represented an acceleration of what Foucault called "statistical medicine, of a medicine of large numbers" within the "state-medical field."[18] The study introduced forecasts, estimates and measurements, across many social phenomena, which were designed to permit the establishment of averages that could then be protected by remedial action against those who fell dangerously far from these norms of prescribed acceptability. The cohort study was thus designed to mark out the location and dimensions of the "security mechanisms" that were seen to be necessary "around the random element inherent in a population of living beings so as to optimise a state of life."[19] In this manner, it was an intimate component of the biopolitical approach that pushed British politics into a data turn from the mid-1960s.

However, all these advantages did not mean that cohort studies were straightforward to set up or run. One issue they faced concerned who would administer the tests, conduct the interviews and gather the data. National cohort studies, where the group was spread across the country, necessitated data gatherers who were also geographically dispersed. The problem was that government did not have access to such suitably qualified people in anything like sufficient numbers. This meant that information would have to be gathered by health visitors, teachers or school nurses, and this, in turn, meant that the information that could be gleaned would be limited to objective observations and measurements, whereas what was being sought was far more complex. The only way to circumvent this problem was, it was recommended, to further break with the traditions of the British government's data enclosure and "rely on co-operation from universities and other professional institutions wherever this can be found."[20] Additionally, it was noted that a cohort study was, in its nature, a long-term project. While it was underway, staff would naturally come and go, therefore to maintain continuity, it was vital that the centre of the project be located in an institution capable of providing the support necessary to guarantee this continuity.[21]

The cohort study was designed to be interdisciplinary, to see people in the round, and this meant that at its core, it involved collecting and connecting data on topics such as educational attainment, health, family background or criminality that had traditionally been harvested and held separately. However, it was suggested that the study could go much further than this and that rather than only conduct field research to collect a unified body of information, it should draw together sets of data that already existed in other filing systems to either supplement the specialised cohort trawls, or possibly to replace them altogether. In this way, the cohort study's designers saw that it could be positioned within the push to increase the use of the "masses of data" that government was, in Donnison's phrase, "sitting on" and by doing so operate in tandem with systems examined later in this chapter.[22]

Thus, although Hammond had noted that a constant stream of data about the cohort's subjects was unobtainable, it was possible to envisage a cohort study that could come close to achieving this by being built on sets of "information routinely collected for administrative purposes," this would, moreover, be the "most economical" way to develop such a dataset.[23] Schools and hospitals held information, as did the registrar general. In demonstrating the sort of actions that ought to be taken to gather population data, Hammond's report described a study underway in Newcastle. This was "collating all the information collected by the various divisions of the local authority social agencies" to create a bank of "information on a child cohort without the subjects of the cohort being aware that they are the object of study."[24] This precedent, of completely circumventing the established norms of confidentiality, was not, despite Hammond's urging, followed by the 1970 national cohort study. Nevertheless, the fact that similar policies would be adopted by subsequent government initiatives, such as the longitudinal study (examined below) was made clear by the British Association for the Advancement of Science, which, less than ten years later, pointed out that within government "administrative and research data are often mixed, and are collected and transferred without the knowledge of the subjects."[25]

2

The second government population research project examined in this chapter is the GHS. Fieldwork for this began in October 1970, and though it would be abolished in 2012, in the 1970s, it was seen as representing a step change in population data gathering. Thus, by 1977, it had built a reputation as "a unique type of survey" that delivered "substantial" benefits and, as such, represented a development that the CSO was clearly proud of.[26] Indeed even before the GHS began work, Moser was keen to stress its importance. In the first (1970) edition of what would become the government's flagship population data publication, the annual *Social Trends*, he wrote that "this survey [the GHS] will provide a major new source of data on the social conditions of this country ... a regular picture of changing social conditions to help the formation and discussion of social policies."[27]

The survey was conducted by interviewing all the adults (over sixteen years of age) in fifteen thousand households, drawn from the census data, over the course of any year with roughly one-quarter of these being interviewed at each of four points throughout each year. This allowed the government to build up a clear picture of trends if not exactly as they unfolded, then certainly in a much more timely way than any other system. This continuity of surveying also allowed the OPCS to identify for follow-up work minority groups, "such as the coloured population," where it had, by 1977, been able to look at some 5,000 people from a total sample size of 200,000.[28] The GHS gathered data on a wide variety of topics such as housing, education, health, occupation, income, smoking and attitudes to careers in the police and the military. Because this was the first time data had been collected on many of these topics, and certainly the first time it had been collected in one survey, the GHS allowed the OPCS to research the relationships between these data points. This type of work could be undertaken because the survey's breadth of coverage allowed these relationships to be observed and analysed. Moreover, because information was gathered by interviewing every adult in a chosen household, it permitted a clear picture of the nature of the household to be drawn. Information gathered in this way, it was argued, was more likely to be accurate than any derived from ad hoc or single-topic surveys, and it was certainly cheaper than these alternatives. Additionally, as was pointed out by the OPCS's chief medical statistician in 1976, the GHS provided qualitative data that allowed the OPCS to analyse needs that were unmet by existing government services. Other forms of surveying could not provide such insights, and it was on this basis that it was argued that the GHS came close to providing the type of data advocated by Acheson.[29]

The Social Surveys Division of the OPCS, which ran the GHS, had always intended that the interviews that gathered the survey's information should "be maintained as an acceptable social contact."[30] However, this ran foul of another of the founding principles of the project. This was that: "it is very much our intention to keep the General Household Survey as an elastic and adaptable instrument."[31] This meant that it was open to including new questions at the request of the departments that would utilise the survey's findings. Initially, those running the GHS had been concerned that it might not be sufficiently used across government and that they "should do all they can to promote awareness."[32] However, they quickly learned that they had nothing to fear in this respect. Departments had by this time (1972) already built up a desire for population data, and once it was available as a source, the GHS questionnaire ballooned in size. This was because, not only were departments keen to make additions to the survey, but they also proved reluctant to having their existing sections edited down to a manageable size.

Thus, it was noted that in 1971, the schedule ran to twenty-eight pages, but in 1973, it was estimated that by 1974, it would be double this length. Pilot studies showed that the 1971 version took forty-five minutes to administer while the revised and extended version took seventy-five minutes. Though this increase was later reversed, while keeping the number of topics covered more or less the same, this had three negative impacts on the survey.[33] First, the field workers became disenchanted with the length of the survey and with the constant changes being made to it and, perhaps more importantly, congruent on the survey's burgeoning size, response rates were falling. These had declined by 2% between the 1971 and 1972 surveys, and it was suggested that: "steps should be taken to stem the fall before the GHS ceased to be a useful instrument."[34] Thirdly, as was noted somewhat tetchily, in 1976, the flexibility of the survey "in terms of quarter to quarter changes in questionnaires is incompatible with speed, comparability of data over time, aggregation of data over a number of years, time necessary for adequate reporting and, above all perhaps, with timely data processing."[35] The CSO thought departmental users of the survey's outputs had unrealistic expectations of "overnight" results. Whether these expectations were realistic or not, whether the collection, collation and publication of the results from the field could have been done differently or more efficiently is not the main point here. Rather this mounting pressure on the GHS, within a few years of its launch, highlights the extent to which this desire for data had come to the fore across government. Thus, in this sense, the fact that the CSO could complain about other departments' willingness to inflate their components of the GHS indicates the overall success of the data turn in government

This can also be seen in the extent to which the GHS's findings were used. By 1974, the CSO found that they had already been "used by more than a dozen departments and major divisions within departments" with the main users being the Department of the Environment, the Department of Health and Social Security (DHSS), and the Scottish Home and Health Department.[36] However, more importantly, this data was not kept within the government's own enclosures. It formed the backbone of a lot of what was published in *Social Trends* (thirty-two tables in the 1976 edition, for example[37]) while both the Post Office and the Greater London Council also utilised the survey's material. But above all, researchers who were outside government were able to use this information. This was not coincidental. In fact, the Social Survey Division of the OPCS maintained that: "this kind of outside use should be encouraged, as it extends the value of the survey far beyond that which could be achieved by departments alone."[38] A remarkable attitude when compared with the one prevalent across government before the mid-1960s and one that marked the GHS as a vital element of the new data regime established at the heart of government. These academic researchers used material from the survey that covered "almost the whole range of variables contained within the GHS," and they were at the forefront of those using this material to tease out interrelationships between its data.[39] For example, Richard Layard of the Higher Education Research Unit of the London School of Economics, strongly supported by Moser, used GHS data on income, schooling and employment to form the basis of a paper on social mobility. Most notable of all Donald Acheson, who had campaigned for access to government data for many years, was able to use GHS data on arthritis and rheumatism among people over sixty-five, to illustrate the differences between the demand for and the availability of health services.[40]

Researchers like Acheson had long-railed against the enclosures that protected government data from linkage, most notable of which were the codes of confidentiality that prevented the disclosure of information. The previous chapter demonstrated how, after the rejection of a new population register, the government's data enthusiasts turned their attention to reforming existing data streams and how this brought them up against these codes. Unsurprisingly, the same events played out with regard to the GHS. Information was gathered for the GHS by interviewers who told their respondents that the answers they gave would be confidential. Redfern and the OPCS defended their professional enclosure and stood by this guarantee, but they came under great pressure to allow the release of data in very small geographical subsets, small enough to possibly allow individual respondents to be identified. Departments argued that the pattern of releasing data only where it applied to populations of 250,000 would not meet their future needs. The OPCS had stated that smaller sets of GHS data could be produced in "exceptional circumstances" and this created a small crack in the confidentiality wall that other departments attempted to widen.[41] The publication of GHS data had been "greeted by wide and enthusiastic acclaim" but after the mid-1960s, with the advent of a new set of attitudes toward the place and importance of population data in government, once data was seen to exist, "problems of confidentiality will persist and new ones will arise."[42]

However, just as important as the way, the GHS made the information it had gathered widely available was the fact that it also sought to disseminate the methods it had successfully used to gather it. The GHS had much better non-response rates (15%) than did either the Family Expenditure Survey (30%) or the National Food Survey (45%).[43] Though its results sometimes threw up anomalies that needed to be ironed out, by comparison with other data sets, this response rate compensated for its relatively small sample size and made its findings more representative and reliable. The Department for the Environment had, for example, put questions into the GHS asking people about the number of cars available to their household and any long-range journeys that people had made. The results for this were significantly at odds with what the ministry knew from other sources; nevertheless, after reviewing these anomalies, it decided that: "it seemed that the GHS was the better source of information."[44] Given these methodological superiorities, the GHS was viewed as an overall success, and, in 1972, the Social Survey Division within the OPCS produced a series of methodological papers for circulation, inside government and within the wider research community, detailing the lessons learned from running the survey in order to stimulate more data gathering.[45]

3

Another of Redfern's data innovations that was adopted, and which, like the cohort study is still (in 2021) running, was the longitudinal study established in 1971. Both the cohort study and the GHS were projects where the government, in the form of the OPCS, went out into society and undertook research into the lives of the population. The longitudinal study, by contrast, provides an example of how government attempted to link data it already held, as a result of its routine operations, into a systematic matrix that it could use to develop wider and deeper knowledge of the population. Chapter 1 demonstrated how medical research was one of the main currents that fed the demand for and expansion of population data in this manner, and that Donald Acheson was at the forefront of this drive for data. Thus, Redfern must have been pleased to be able to quote Acheson's view that the longitudinal study had the potential to be "the most significant development in the field of routine social and medical statistics since universal registration of births and deaths became mandatory in 1837."[46] This section appraises the nature, role and significance of this development.

Like cohort studies, a longitudinal study was not something novel in the 1970s. Such a study had been suggested by the second report of the North Committee in 1950, the SMU always saw this type of work as being its raison d'etre and, in 1969, the OPCS's medical statistician suggested something similar.[47] Longitudinal studies were defined, by a progress report issued by the CSO's sub-committee overseeing the project, as being "not so much concerned with the collection of data on characteristics of a population, as with the testing and discovery of causal hypotheses."[48] As such, like cohort studies, they needed researchers with specialised skills and institutional support, "interdepartmental machinery," to be maintained for a long period.[49] Previous longitudinal studies had been "essentially ad hoc and guided by short-term considerations," but this study was designed from the beginning to become a fixed point of the government's new data-collection systems.[50]

This study was initiated in March 1971 when Redfern wrote to Sir Keith Joseph, the Conservative government's secretary of state for health and social security, requesting permission to set it up. He received this permission and the study was duly established. Redfern's letter to Joseph highlights the pressure the British political system continued to exert on those who sought to increase the flow of population data, to conform to, or at least to acknowledge the traditional British norms of data confidentiality. The letter opened by reassuring Joseph that the planned scheme could be implemented "without the slightest risk of infringing confidentiality," and closed by re-emphasising "that there is no cause for public apprehension on the privacy front."[51] Seven years later, a paper giving an outline history of the study similarly stressed how "confidentiality considerations have been kept very much in mind."[52] The study drew its sample of the population (a total of 548,539 individuals) from people whose births were registered on four days of each year. It was decided from the beginning of the study that these dates would be kept secret and would only ever be revealed were they to be the subject of parliamentary questions. But even were this to happen, the study was built around the fact that the dates selected would change each year thus preserving the anonymity of members of the sample.

In addition to this stress on confidentiality, Redfern's letter to Joseph also highlighted four features of the longitudinal study that made it so attractive to the government's population data enthusiasts. The first of these also allowed its proponents to argue that the study was designed to protect confidentiality since it did not identify its subjects by name, but rather by a number assigned to them. This was necessary because in essence, the whole project existed to link data currently held in two separate data silos in mutually incomprehensible formats. These were the National Health Service Central Register (NHSCR), which included data on the vast majority of the population, and the census. Subjects were selected by date of birth from the census data and then flagged up on the NHSCR. The OPCS would use this latter source to build up a picture of life events occurring to the subjects that were recorded in the health system, such as births, deaths and cancer registration and, assuming people re-registered with a doctor on changing address, it would also allow the OPCS to build a picture of domestic population movements. While the census data for its part would allow the study to record migration and marriage. None of this could have been done without identifying the subjects in a standardised way and this was achieved by giving each of them a unique eight-digit number. It was these numbers that provided Redfern with a means of assuring ministers that the study would respect its subjects' confidentiality, which is ironic since elsewhere, as has been seen, Redfern was at the forefront of those pressing for the adoption of such numbering schemes to link different datasets and so wear down the norms of British confidentiality. Indeed, in many senses as will be seen, the linkage on which the longitudinal study was based was, even in the 1970s, becoming the norm of population-data gathering.

The second aspect of the study that made it so attractive to those driving the data turn in government, was the outcome of this linkage. This study achieved what had been seen as the aim of data linking since Acheson's Oxford study: it aimed to see people more in the round. This was a phrase widely used to describe the functioning of the cohort study (above) but where the cohort study involved collecting and collating data through research, the longitudinal study achieved the same aim by linking data that had always been held by government departments but which was held apart by their codes of confidentiality. Through its linkage schemes, the project focused on finding interrelationships.[53] For example, the study aimed to cross reference the existing cancer records to indexes of social class (from the census data) to permit a more thorough analysis of cancer in society.[54] In doing this, the longitudinal study was part of the data zeitgeist that this study both accepted and accelerated by responding to the requests it received from across Whitehall for data tabulated to illustrate these interconnections.[55]

The research-based studies examined above, the cohort study and the GHS, involved either expensive and time-consuming interviews with their subjects, or devising, distributing and assessing questionnaires sent to them. However, the third feature of this longitudinal study that made it so appealing to Redfern was that it linked, and so sought to increase the use of data that was already in the hands of the OPCS as "a by product of other OPCS procedures."[56] The study therefore simultaneously increased the efficiency of data operations while also obviating the need to pester the public by asking them to provide more information. The project would, therefore, boost the efficiency of government data holdings, "an expensive investment from which we have not, in the past, been able to draw full dividends."[57] Consequently, it would allow the GSS to more fully utilise "the data capital" that was its "prime resource."[58] Thus, as the CSO noted, "the marginal cost of using it will be trivial in comparison to the costs of a special survey."[59] Indeed, in writing to Joseph, Redfern encapsulated the entire exercise by stating that it consisted of: "no more than arranging into more usable form data we already hold on one in every hundred of the population."[60] The longitudinal study is thus an example of the system noted by the British Association for the Advancement of Science (above) whereby "data are often mixed ... without the knowledge of the subjects."[61]

In addition to these three points, there was another reason why the study was so popular with the government's data gatherers. This was the fact that it allowed them to validate and hone their statistical procedures. Thus, in 1984, a paper presented to a workshop on the longitudinal study at City University noted that the relationships observed through the study, "have led to a further understanding of the nature and limitations of the different data sets held in OPCS and of the LS [longitudinal study] data themselves."[62] This validation process had been built into the study from its inception. Indeed, when the study was first mooted, in 1969, its main purpose was envisaged as being the correction of discrepancies in measures of national mortality rates that occurred because of differences in how any individual's occupation could be recorded by the census and on a death certificate. The longitudinal study (LS), it was suggested at that point, would enable "truer occupational mortality estimates to be obtained."[63] As was seen in the previous chapter, the pursuit of ever wider and deeper knowledge of the population was an inherent feature of the government processes started under Wilson in the 1960s, and the government's data collectors were, obviously, at the forefront of this. These people, such as Redfern, not only saw the overall benefit to society of what they did, they also had a professional and occupational stake in these processes. Thus, while they ultimately sought true data, one step along the route to this goal was to get the type of truer data that the LS seemed to promise. It was in this vein that a 1980 review of the Scottish version of the study (Scotland had its own registrar general and so separate sets of population/census records) noted that one of the big advantages of the whole study was that it would "give information as to the effectiveness of the linkage methods used."[64] The only way to collect better data was, it seemed, to practice by collecting data.

Nevertheless, ever impatient for improvements, Redfern also pointed out that the study still had shortcomings that would benefit from further refinements. Here he highlighted its "linkage mechanism." This was inadequate, he argued, because it "depends on unreliable identifiers and on checks of identities against the NHSCR."[65] As a result, the rate of successful linkage for the surveyed population was only 91%. It is worth noting that this level of linkage between two completely different datasets would have been seen as phenomenal in the early 1960s and equally would have been regarded as a success when the LS was launched in 1971, but here, in an article comparing British systems to those used abroad, this level of linkage across datasets was seen as yet further evidence of the "ramshackle" nature of British population data.[66]

This desire to improve operational skills and expand surveying capacity was not stated in Redfern's initial letter to Joseph, but once the study was in place, it established a methodological precedent and was used as such. Thus, in 1983, the OPCS wanted to establish a second national morbidity study, the previous such study had attracted hostile publicity and those seeking to implement the later version were keen to avoid this. In doing so, Redfern made the case that it was "proper" to link the sets of records concerned "providing that methods to safeguard confidentiality would be akin to those adopted for the longitudinal study which itself had created a precedent for the linkage of records, including census records, within the office."[67] When it came to expanding the government's population data holdings, or to enhancing their standardisation, centralisation or linkage, once a precedent had been set, it would be followed.

4

In August 1978, in his final report on the condition of the GSS, Claus Moser argued that the organisation's purpose was primarily "to provide a service that will be used by government." He argued how, while fulfilling this mission under his leadership, the CSO had boosted its role in coordinating government statistics. However, he added that this had not progressed as far as he would have liked due to the "lack of both a central social policy and a conceptual framework in social statistics."[68] In other words, the retiring chief statistical officer argued, like Foucault, that power "and statistics mutually condition each other."[69] However, as Moser would have been well aware, the absence of such a central driver of social policy did not mean that the government had never sought to introduce one. In fact, in the 1970s, a centralised social policy initiative was launched by the Central Policy Review Staff (CPRS), which had been set up under Edward Heath and tasked with modernising government (the CPRS was led by (Lord) Victor Rothschild and would be abolished under Margaret Thatcher in 1983.) This policy initiative began as a drive for overall centralisation and rationalisation across Whitehall, but was later streamlined into the JASP. Under Wilson (back in office after the election of February 1974), the JASP was launched in August 1975. But the feasibility studies that would have realised its vision were spiked by the economic crisis and this, along with departmental hostility to central control of policy, meant that although it was launched, the JASP never actually went into operation.[70] However, in the same way that Wilson's thwarted aim of introducing a population register (the *People and Numbers* plan), in the late 1960s, made the government's overall aims clear, so an examination of the aims of the, similarly stymied, JASP can allow further charting of the contours of the government's thinking about population data and highlight the centrality of plans for data linkage to this thinking.

In Chapter 1, it was shown how from the immediate post-war period, social scientists and medical researchers argued for data on ostensibly different aspects of life, such as health and the economy, to be linked. This was the view that underpinned the departmental merger that created the DHSS, and it also underpinned everything done by the LS. The JASP sought to use this policy vector of linkage to understand what David Donnison described as: "the Pandora's box of urban problems, social, economic and political," which is to say that research was vital to understand society and so form better, more effective, policy.[71]

The ministerial report on the JASP was finalised by June 1975. Its overarching aim was to drive social policy further and faster in the direction of linked policy to address issues in people's lives that overlapped departmental boundaries. In short, the JASP wanted "improved co-ordination between services as they affect the individual."[72] Ministers in the Labour government wholeheartedly agreed, offering their "strong support" for the "CPRS's approach: social problems interlocked."[73] One important feature of the JASP was that it promoted centralisation. Indeed the report was quite clear that policy trends promoting local autonomy were "not easily compatible with continuing attempts by central government to ensure that specific problems are given high priority," and thus concluded that: "there may well be a case for doing more to limit variations, for example by monitoring standards."[74] Wilson chaired the ministerial group set up to steer the JASP and here is it noteworthy that, at the first meeting of this group, there was agreement that the CPRS's next project should be an examination of the relations between central and local government. However, the biggest factor behind the centralisation inherent in the JASP was the fact that it was an exercise in data use and, as has been seen, the government's increased use of data was a byword for overall centralisation.

The report opened by listing the factors that impinged on a government's ability to deliver its manifesto commitments, which included economic constraints and the legislative timetable. But the factor on which the report centred was "a serious lack of information about many social problems," which left the government with "no real basis for assessing need or the effectiveness of provision."[75] The government lacked systematic thinking about social policy because it lacked the information that could feed such thinking. For example, the DHSS told the CPRS that: "we undoubtedly need more information about the needs of our groups, the geographical distribution of those needs and the extent to which they are being met."[76] Connections between social policy makers and statisticians were in need of strengthening and the work of the diffuse GSS needed more tight coordination. Policy makers ignored, downplayed or misunderstood the vital role of research and information, which consequently tended to be neglected, thus fuelling the status quo.[77] Were data to receive the backing it merited, the report argued, the system could be equipped to act preventatively to stop problems from arising in the first place and thus allow resources to be targeted more effectively.

This language, of targeting scarce resources, was particularly important by the mid-1970s. Thus, in June 1975, when he was addressing the National Council of Social Services, Wilson repeatedly told those present that government budgets were tight and that: "as I have said in government, we must face the fact that resources will, over the coming years, be severely limited,"[78] all of which parsimony was occurring at just the time when expenditure on social service was rising often to address issues uncovered by researchers, such as loneliness in old age and juvenile delinquency. Having made these points, Wilson then read two pages of the JASP report to those assembled in order to underline how a joint approach could allow social services to meet real needs even though resources were limited.

As a result of this thinking, all the report's recommendations concerned improvements to the government's data flows. These included the establishment of a group of senior CSO social statisticians; more effective presentation of statistical information; the inclusion, in all spending proposals, of an analysis of the wealth distribution impacts of the proposal; and five research projects examining how policies and institutional provision intersected and effected particular groups of people. All of which was encapsulated in the phrase, "social monitoring," used in the report, which the CPRS acknowledged it had taken from Anthony Crosland.[79] Indeed so central was social monitoring to the whole enterprise that when ministers considered the report before it was published, the point was strongly made to Wilson that the CPRS had been too timid in making their case. Social monitoring was "the most important proposal discussed" and the whole project would stand or fall on how this was implemented. It was, they argued, "altogether unsatisfactory" that decisions were made without any data to determine their impact on wealth distribution. As a result, they argued that: "we should, as Crosland suggested, get on with it and make a start now."[80]

In 1973, the CPRS was aware that it needed to locate its work in relation to other inquiries. In doing so, the CPRS demonstrated that it was more than willing to follow its own lead and was keen to use any data it could access across government to further its initiatives. Thus, it made extensive use of GHS data in developing a data matrix to delineate the main client groups for the DHSS.[81] However, the GHS was not the only data-gathering project underway.

The CSO supported the CPRS's work, and in May 1973, just as the CPRS was starting the work that would produce the JASP proposals, the CSO finalised a programme analysis and review (PAR) on statistics for social policy to underline the key role it could play in the JASP. This was designed to study the possibility of drawing together data "by linking samples of existing administrative records which are normally held separately."[82] It was begun on the authorisation of the Conservative prime minister, Edward Heath, and was due to begin with a year-long feasibility study. This study was to range over all of Whitehall and was designed to survey how data could be linked within and between departments and to receive and make suggestions for this.[83] This was seen as necessary since "in some departments, thinking of this kind may be precluded or inhibited by existing restrictions (statutory or established by long precedent) on the handling and the transfer of administrative data."[84] In other words, the purpose of the study was to design methods to circumvent the barriers posed to the overall linking of data by "confidentiality issues and so on."[85] Thus, it was envisaged that the study would be followed by a review of the legal, administrative and political implications of constructing a pan-Whitehall data web. Though the CSO was keen for its PAR to be associated with the JASP, it was also keen to pursue its own agenda and to maintain its independence from the CPRS project. Thus, it was argued that the JASP would look at the possibility of developing new sources of data, whereas the CSO's administrative data project would, as its name implied, focus on increasing the usage of data that was already on file. Moser and the CSO seemed to be aware that though what the CPRS was proposing was, in data terms, similar to its own work, in political terms, by seeking to centralise policy itself rather than just the data to feed policy, the CPRS was trying to take a bold step.

In fact, a few weeks before the official launch of the JASP, Moser briefed CSO staff that the CPRS's proposals on social monitoring "are to be treated as tentative."[86] Moser was aware these proposals would pose considerable administrative and political problems, as their interdepartmental focus would be correctly seen as part of a drive to centralise control of not only data, which of course he both wanted and knew was difficult in itself, but also policy. His strategic political awareness proved to be correct and the JASP was never adopted across government and was largely abandoned by 1977. William Plowden, who was heavily involved in the CPRS, noted that it failed because it could not make "headway against departments which in the view of one official had 'ganged up' against the CPRS, nor against the weight of the vested interests, such as the professional associations, in both local and central government."[87] The CSO's support for the JASP can be seen as another attempt by the government's data enthusiasts to engage in a war of manoeuvre against the enclosures that prevented the free circulation of data. Moreover, with hindsight, it is possible to see that, judged from their perspective, Moser and the CSO acted shrewdly in keeping their component of the JASP as a separate entity.

The CSO's PAR on statistics for social policy developed themes that the CSO had attempted to pursue before the advent of the JASP, and which survived its downfall. Moreover, though it was set up under a Conservative government, it was able to survive a change of governing party since it was based on a principle that appealed across the party divide. This insisted that data about people that restricted the gaze to only one facet of their lives, or to one relationship they had with a government body, for example, viewing a person as a recipient of a particular benefit, was "of limited use for many policy purposes." Understanding the contents of the Pandora's box of social policy required, the PAR maintained, "data which observes people 'in the round.'"[88] This language was virtually identical to that used by Labour thinkers on the same subject. Thus, in 1965, Tony Benn had written that government processes, which took "a narrow view of human beings," whereby "the agencies that deal with us and our problems are too often concerned only with the narrow sector of our lives which touch their particular field," were "both destructive of the individual personality and also unscientific."[89] Similarly Anthony Crosland noted that once in power, Labour would "increasingly need to focus attention, not on universal categories, but on individual persons and families" and that targeting government aid would need the guidance of "patient, empirical social research."[90] Moreover, in 1972, when addressing the Royal Statistical Society in his capacity as its president, Wilson referred to an example of such research that had been conducted by the MP for the Gorbals with the aid of student volunteers from Glasgow University. This had revealed, in "horrifying" detail, "the dimension of the legion of the forgotten." From this, Wilson asked his audience to draw the conclusion that "the task of the social services, Government, local authority and charitable bodies, will be not to measure numbers, but to identify individuals."[91] Addressing another gathering of statisticians three years earlier, Wilson had told his audience that: "our objective today is to bring about compassionate as well as efficient societies and the statistician has a very important contribution to make towards the achievement of both."[92] Thus, by the early 1970s, there was a consensus across the party divide on the broad nature of social policies and pointedly on the fact that these policies needed to be driven by a centralised and linked pool of population data.

The main purpose of the PAR was to fulfil these interdepartmental needs for statistics for social policy and, in seeking to do this, it broke this generic goal down into four specific policy objectives. These were, first, the identification of what data was needed for interdepartmental policy to progress, getting this to policy makers in time and commensurate with this, not pursuing projects that served no real purpose. Second, in order to "raise ourselves above already accepted ideas," it was necessary to relate all data collected to the policies that it was supposed to serve, and weed out streams that did not meet these requirements. Third, all data gathered, wherever it was gathered, should be disseminated as widely as possible across Whitehall. Fourth, it was argued that because data and the institutions collecting it were heavily inter-related, it was necessary to have a review of the structures of the GSS because: "if the GSS were organised on a centralised basis for example then co-ordination would be quite a different task."[93]

In times of austerity, an argument that was commonplace amongst the advocates of enhancing government population data flows carried extra weight. This line of reasoning held that since data was expensive to obtain, once it had been gathered and filed, it ought to be used as often and as widely as possible. Codes of confidentiality and professional enclosures prevented such cost-effective usage of resources of course, but so did the fact that different departments held their data in different systems and filed information using different referencing systems, all of which meant that the government's system was (to borrow Blair's phrase) an archipelago of data islands rather than a contiguous body. The key device brought forward to increase and ease navigation between these outposts was the use of a common numbering system. People's names could not be used to link different datasets in this way, as they did not provide a set point around which Whitehall could navigate. Names were flexible: the order they were used in could be altered, spellings could vary and names were often abbreviated. What was needed, the argument ran, was for people to be identified by a number that would remain unchanged on all their files for life. George North had suggested that the war-time National Register, had it been preserved into the peace, could have provided this numbering system, and the topic was an integral element of the *People and Numbers* plan that was shelved by Wilson in September 1969. Indeed, the idea was a common theme in discussions among the government's data gatherers. Thus, at a November 1972 meeting of the CSO's Committee on Statistics for Social Policy, Moser stressed the importance of the standardisation and comparability of data and argued that: "much would be lost without the facility of unique identification."[94] As was seen in the last chapter, Moser and Redfern had differences of opinion about the operation of the OPCS, but they spoke with one voice on this subject of common numbering to facilitate the greater use of administrative data. At the point when the CSO's study of the use of administrative data for statistical purposes was being readied, Redfern noted that this use of common numbering was "an obvious reform" and that once underway, the study would inevitably need to consider whether these common numbers should be drawn together in a population register.[95]

One reason why Redfern could argue that this was an obvious step forward lay in the fact that he could point to other states where this was done. In fact, just as British governments would often express exasperation at the paucity of the data resources they had at hand to fulfil the tasks they wanted to undertake, so they would often comment covetously about the resources available to other states. Thus, in 1962, members of the Statistics Department of the Board of Trade, visited the EEC Statistical Office in Brussels and, in reporting on this visit, they gave two reasons why they approved of what they had seen there. First, the Treaty of Rome, they found, was less onerous than might be imagined in Britain, it gave little scope for central institutions to impose demands on member states and thus central data collection worked almost exclusively by persuasion with little or no coercion by the centre. Second, this system produced no fewer than ten published statistical series, "a quite formidable output" and one that plainly surpassed its British equivalents.[96] The position occupied in the minds of British government data gatherers by European institutions would later shift from this one, of Europe being a beacon to follow, to one where Europe was Britain's data bête noire (see Chapters 5 and 9), but some states would always be held up as exemplars to emulate. Prime among these were Israel, France and particularly Sweden.

One academic author made a comparison between British and Swedish record-keeping and noted that: "incredible as it might seem to most people," British government departments did not link their records and so lagged lamentably far behind Sweden.[97] This incredulity was also felt across government. Thus, *Proposals for a Population Register System in Great Britain* was the title of a paper written from the General Register Office, probably by the registrar general, Michael Reed, in October 1967.[98] A British delegation had attended a symposium on population registration in Jerusalem and the paper was written as a direct result of how impressed its members had been by what they had learned about the Israeli and Swedish registration systems, and this became one of the sources of information and pressure that led to *People and Numbers*.[99] In 1978, Moser noted that more centralised systems (his example was the French) had a more coherent political identity and were protected by laws that allowed them to receive, collate and deploy a wider variety of data than was the case in the GSS. He added that a move to such a system "should be possible here" and, were it to be achieved, he looked forward to "the statistics divisions receiving anonymised but linkable data and producing new analyses on behalf of the statistical service."[100] Such a comparison, with "the more advanced practice abroad," was suggested as part of the 1973 PAR on social statistics, and similarly, in 1978, an examination of the GSS's data management requirements cast envious eyes over the Swedish system. Here, the report found, the basic premise of all data holdings was that: "it should be possible to increase the return from the data capital by storing primary data in such a way that it could easily be re-used for purposes other than those originally intended."[101] Moreover, this review found that of the thirty-six foreign systems it examined, thirty-three had similar requirements to the Swedish.[102] In other words, Sweden, and a lot of other states too, already had exactly what the British government wanted and what the 1973 PAR clearly sought to inaugurate in Britain.

During the period covered by this book, the depth and breadth of the British government's data operations expanded, but the British government never stopped comparing itself unfavourably to others. This was the case even when the British started to put world-leading systems in place. The CSO wrote that the GHS was seen as "very different from existing multi-purpose surveys in other countries and there are now indications that others may wish to follow the lead given by this country," adding that these others included Sweden, which "is to start a similar survey this autumn."[103] But despite developments such as this, in 2002, the Blair government justified its changes to the registration of births, marriages and deaths, by arguing that its proposed central database would be similar to those used abroad (Canada and Australia were the examples given).[104] A few months later, this government published a study on identity fraud that exposed how far in advance of Britain some selected foreign examples (the United States, Spain and Germany were chosen for this purpose) were in tackling this problem.[105] The British government saw improved data flows as being the key component of modernity, the one that would allow access to the population and so facilitate the biopolitical interventions that characterise modern government. The problem the British government had in this regard was that, just as attempting to know the population put it in pursuit of a chimera, so this constant comparison with other states would ensure that the British government's data enthusiasts would always find their own systems wanting as those of their chosen international comparators were also constantly changing. Put simply, there never was, and never can be, an objective yardstick of modernity.

Thus, it should be acknowledged that these international comparisons were never intended to be simply descriptive rather they were always a call to action. Redfern may have argued that an examination of the potential for the common numbering of government files in Britain was an obvious step along a road well trodden by others. But rather than simply assume that the general run of events would lead the study in this direction, he drew up the plans necessary to spur this move. These plans were put to the cabinet's Home Affairs Committee in November 1975 and were all of a piece with the thrust of the government's data drive in general and the nature of programmes such as the JASP and the CSO's PAR on social statistics in particular. Thus, because his focus here was on increasing the use of data that already existed, he wanted to set up a feasibility study on extending the electoral register in such a way that it could be used to produce more complete population data.

Stella Cunliffe, the head of statistics at the Home Office (she would later become the RSS's first female president), presented her view of OPCS's case and began by agreeing with the commonly held view that if data was not linked across departments then either valuable information would not be available to policy makers or it would need to be collected on multiple occasions with all the expense that this involved. In this, she was in broad agreement with Redfern. But the OPCS plan, she wrote, aimed to set up a population-record system run through the electoral register, and to do this, the OPCS would assume control of the registers. To achieve this control, she added, the OPCS wanted to set up a permanent staff at ninety offices spread across the country, the cost of which, she pointed out, had not been included in the proposal. Furthermore, in her opinion, this vast extension of the OPCS and the use of the electoral register were unnecessary since the type of data it might provide could be obtained by sampling the population.[106] The General Department of the Home Office concurred with Cunliffe's view in recognising the usefulness of this data but argued that the electoral register was not the right place to source this information. People, it argued, provided information to electoral officers in return for the franchise. The data provided was used for this purpose alone and it was on this basis that the public accepted these registers. Any decision to broaden the use of this information would be one to which the Home Office would be "in principle ... strongly opposed."[107]

Redfern's proposals to extend and mine the electoral register for population data were defeated at this juncture, but what is noticeable about this incident is that there was broad agreement, even amongst those opposed to this tactic, that it was engendered by the correct strategy. Thus, despite the fact that it may appear as though the Home Office was, in this case, standing foursquare behind the data traditions of Britishness, the reality was more complex with the need for population data being universally accepted across government. On this basis, though it may have been defeated in 1975, this plan would resurface within a few years (see Chapter 6) while the principles that underpinned this, the widely recognised need for a pan-Whitehall data network constructed around common numbering, would always be at the centre of the plans drawn up by the government's data enthusiasts.

Indeed the comments on the Swedish system (above) were drawn from a 1978 CSO report on the computer requirements of the GSS. This noted that the ultimate purpose of the GSS was to provide the government with up-to-date information and that to do this, it needed easier access to data. This, it argued, could be provided if data were more compatible across departments, and this, in turn, could be achieved by collating data within departments. These departmental packages could then be drawn together if the civil service as a whole adopted "common and standard referencing systems."[108] This standardisation and linkage thus mapped out a route that could circumvent the perceived problems within the existing system, one that was inflexible, uncoordinated and lacking in standardisation, and where data was therefore inaccessible. It would allow Britain's population data systems to modernise to at least the level of Sweden's, and it would allow the government to more fully exploit the population data that it held: material that was its "prime resource."[109] It would, in short, learn the lessons of these trials and systems and thus realise Wilson's data revolution.

Once started this revolution almost by definition had no predetermined end point or strategic destination because perfect population data was a chimera. Nevertheless, many of the tactical goals pursued by the, predominantly Labour Party, governments in the 1964–79 period would be achieved. That some of these would be finalised by subsequent Conservative governments speaks to the growing and central importance of data to government as such, rather than to government of one or other party. Thus, while these later governments would for sure use population data to implement different policies, what is equally sure was that they would extend the data they held on the British population. These points of difference and continuity are examined in the next chapter.

Notes

1 TNA, CAB 108/782, CSO, Committee on Statistics for Social Policy, Sub-Committee on Longitudinal Study, *OPCS Longitudinal Study, Note by the Secretary*, 3.

2 TNA, CAB 139/742, *Draft Speech for the Royal Society of Statisticians Banquet on the Occasion of the 37th Session of the International Statistical Institute*, 10 September 10, 1969, 12.

3 TNA, PRO 30/87/61, David Donnison, *The Contribution of Research to Social Policies and Programmes*, Feb. 1972, 18.

4 TNA, PRO 30/87/511, David Donnison, "The Age of Innocence is Past: Some Ideas about Urban Research and Planning," *Urban Studies* 12 (1975): 271.

5 TNA, BN 29/1820, Home Office Research Advisory Committee, Dr. W. H. Hammond, *The Place of Cohort Studies in Social Research*, n.d., ca. Dec. 1965, 1.

6 TNA, BN 29/1820, Home Office Research Advisory Committee, Dr. W. D. Wall, *Cohort Studies*, n.d., ca. Dec. 1965, 1.

7 Ibid.

8 Ibid., 1-2.

9 TNA, BN 29/1820, Home Office Research Advisory Committee, Hammond, *The Place of Cohort*, n.d., ca. Dec. 1965, 10.

10 TNA, BN 29/1820, Home Office Research Advisory Committee, Wall, *Cohort Studies*, n.d., ca. Dec. 1965, 3.

11 Ibid., 1.
12 Michel Foucault, *Psychiatric Power* (New York: Picador, 2006), 77.
13 TNA, BN 29/1820, Home Office Research Advisory Committee, Wall, *Cohort Studies*, n.d., ca. Dec. 1965, 2–3.
14 Michel Foucault, *Abnormal* (London: Verso, 2016), 25, and Michel Foucault, *Society Must Be Defended* (London: Penguin, 2004), 247.
15 TNA, BN 29/1820, National Bureau for Co-operation in Child Care, Home Office Research Advisory Committee, *Cohort Studies*, May 1966, 2.
16 TNA, FD 1/286, Medical Research Council, *Social Medicine Research Unit, Report*, 1948–51, 1.
17 TNA, BN 29/1820, Home Office Research Advisory Committee, Hammond, *The Place of Cohort Studies*, n.d., ca. Dec. 1965, 4.
18 Foucault, *Psychiatric Power*, 248 and 96.
19 Foucault, *Society Must Be Defended*, 246.
20 TNA, BN 29/1820, Home Office Research Advisory Committee, Hammond, *The Place of Cohort Studies*, n.d., ca. Dec. 1965, 3 and 10.
21 Ibid., 3.
22 TNA, PRO 30/87/511, Donnison, "The Age of Innocence is Past," 271.
23 TNA, BN 29/1820, Home Office Research Advisory Committee, Hammond, *The Place of Cohort Studies*, n.d., ca. Dec. 1965, 5.
24 Ibid.
25 British Association Study Group, "Does Research Threaten Privacy or Does Privacy Threaten Research?," in *Censuses, Surveys and Privacy*, ed. Martin Bulmer (London: MacMillan, 1979), 52.
26 TNA, CAB 108/774, CSO, CSO, Committee for Statistics on Social Policy, Working Party on General Household Survey, *Report of the Review of the General Household Survey*, 27 Apr. 1977, 3; and TNA, CAB 108/775, M. J. Murphy to Eric J. Thompson, 27 Oct. 1977, 3.
27 TNA, CAB 108/773, CSO, Committee for Statistics on Social Policy, Working Party on General Household Survey, *Supplementary Uses of the GHS, Annex 2*, reprint of: Sir Claus Moser, "Some General Developments in Social Statistics," *Social Trends* (London: H.M.S.O., 1970).
28 NA, CAB 108/775, Murphy to Thompson, 27 Oct. 1977, 2.
29 Abraham Manie Adelstein, "Policies of the Office for Population Censuses and Surveys: Philosophy and Constraints," *British Journal of Preventive Social Medicine* 30 (1976): 5.
30 TNA, CAB 108/772, CSO, Committee for Statistics on Social Policy, Working Party on General Household Survey, *Questionnaire Review*, 5 Apr. 1972, 1.
31 Ibid.
32 TNA, CAB 108/773, CSO, Committee for Statistics on Social Policy, Working Party on General Household Survey, *Meeting, 10 Feb. 1972*, 7.
33 TNA, CAB 108/773, CSO, Committee for Statistics on Social Policy, Working Party on General Household Survey, *The Working of the G.H.S. Since Its Inception*, 26 Sept. 1974, 145.
34 TNA, CAB 108/773, CSO, Committee for Statistics on Social Policy, Working Party on General Household Survey, *Meeting, 26 Nov. 1973*, 1.
35 TNA, CAB 108/774, CSO, Committee for Statistics on Social Policy, Working Party on General Household Survey, *The General Household Survey 1971–1975, A Survey Surveyed*, 23 Mar. 1976, 10.
36 TNA, CAB 108/773, CSO, Working Party on GHS, *The Working of the GHS*, 26 Sept. 1974, 4; and TNA, CAB 108/773, CSO, Committee for Statistics on Social Policy, Working Party on General Household Survey, *Uses of the General Household Survey*, 17 Apr. 1974, 1.

37 TNA, CAB 108/774, CSO, *Report of the Review of the GHS*, 27 Apr. 1977, 12.
38 TNA, CAB 108/774, CSO, Working Party on GHS, *The GHS 1971–1975; A Survey Surveyed*, 23 Mar. 1976, 5.
39 TNA, CAB 108/774, CSO, *Report of the Review of the GHS*, 27 Apr. 1977, 12; and TNA, CAB 108/774, CSO, Working Party on GHS, *The GHS 1971–1975, A Survey Surveyed*, 23 Mar. 1976, 7.
40 TNA, CAB 108/773, Barnes to Roberts, 10 Apr. 1974, 2.
41 TNA, CAB 108/772, CSO, Working Party on the GHS, *Meeting*, 10 Feb. 1972, 3–6.
42 TNA, BN 102/8, to: Secretary of State, *Problems of Confidentiality and Personal Records*, Apr. 1972, 1.
43 TNA, CAB 108/769, CSO Committee for Statistics for Social Policy, Working Party on the General Household Survey, *Validation of Statistical Output from the General Household Survey*, 14 Jan. 1971, 3.
44 TNA, CAB 108/774, CSO, Committee for Statistics on Social Policy, Working Party on General Household Survey, *Meeting 18 Sept.* 1972, 7.
45 Ibid., 9.
46 Quoted in, Philip Redfern, "Population Registers: Some Administrative and Statistical Pros and Cons," *Journal of the Royal Statistical Society. Series A (Statistics in Society)* 152 (1989): 7.
47 TNA, RG 19/863, *Paper on Setting up and Operation of the Longitudinal Study*, 28 Feb. 1978, 1.
48 TNA, CAB 108/782, CSO, *OPCS Longitudinal Study, Note by the Secretary*, 27 Sept. 1973, 3.
49 TNA, CAB 108/782, CSO, *Committee on Statistics for Social Policy, Sub-Committee on Longitudinal Studies, The Tasks of the Sub-Committee, Note by the CSO*, 1 Dec. 1972, 1–2.
50 Ibid., 3.
51 TNA, RG 19/863, Redfern to Joseph 18 Mar. 1971, in, *Paper on Setting up and Operation of the Longitudinal Study, Annex A*, 28 Feb. 1978, 57 and 60.
52 TNA, RG 19/863, *Paper on Setting up and Operation of the Longitudinal Study*, 28 Feb. 1978, 5.
53 Ibid., 4.
54 Ibid., 31.
55 Ibid., 44.
56 Ibid., 10.
57 TNA, RG 19/863, Redfern to Joseph 18 Mar. 1971, 57.
58 TNA, BN 89/204, *PAR on Social Statistics*, 14 May 1973, 7; and TNA, CAB 108/417, CSO, *Report of the Working Party on the Statistical Service's Data Management Requirements*, 29 Nov. 1978, 8.
59 TNA, CAB 108/782, CSO, *OPCS Longitudinal Study, Note by the Secretary*, 27 Sept. 1973, 2.
60 TNA, RG 19/863, Redfern to Joseph 18 Mar. 1971, 60.
61 British Association Study Group, "Does Research Threaten Privacy?," 52.
62 TNS, RG 19/919, A. Brown and J. Fox, *OPCS Longitudinal Study, 1981 Census/ LS Link, Workshop Paper 1*, 23 July 1984, 5.
63 TNA, RG 19/863, *Paper on Setting up and Operation of the Longitudinal Study*, 28 Feb. 1978, 2.
64 TNS, PREM 9/9, *Report of the Review of Statistical Services in General Register Office (Scotland)*, June 1980, 8.
65 Philip Redfern, "Source of Population Statistics: An International Perspective," *Population Projections: Trends, Methods and Uses* (London: OPCS Occasional Paper 38, 1990), 112.
66 Ibid.

67 TNA, RG 19/864, Dececco to Taylor, 1 Aug. 1983, 1.
68 TNA, PREM 16/1884, Moser, *Review of the Government Statistical Service*, 4 Aug. 1978, 6 and 14.
69 Michel Foucault, *Security, Territory, Population* (New York: Picador, 2004), 317.
70 TNA, HO 411/59, Kent to Shuffray, 29 Mar. 1976, 1.
71 TNA, PRO 30/87/511, Donnison, "The Age of Innocence is Past," 268.
72 TNA, T 227/4395, *Joint Approach to Social Policy: An Initial Action Programme, Draft Paper for Ministers*, 7 Mar. 1975, 3.
73 TNA, T 227/4396, Ministerial Group on a Joint Approach to Social Policies, *First Meeting*, 13 May 1975, 2.
74 TNA, T 227/4395, *JASP: An Initial Action Programme*, 7 Mar. 1975, 6.
75 Ibid., 1.
76 TNA, BN 89/318, *Joint Approach to Social Policy: A Discussion Paper by the CPRS*, 7 Jan. 1975, 3.
77 TNA, T 227/4395, *JASP: An Initial Action Programme*, 7 Mar. 1975, 5.
78 TNA, CAB 184/242, *Prime Minister's Speech, National Council of Social Services*, 27 June 1975, 4.
79 TNA, T 227/4395, *JASP: An Initial Action Programme*, 7 Mar. 1975, 20. See also, TNA, BN 89/318, *JASP: A Discussion Paper by the CPRS*, 7 Jan. 1975, 8.
80 TNA, PREM 16/429, BD to PM, Misc 78, *Joint Approach to Social Policy*, 12 May 1975, 1–2.
81 TNA, CAB 108/773, Barnes to Roberts, 10 Apr. 1974, 1.
82 TNA, BN 89/204, *PAR on Social Statistics*, 14 May 1973, 8.
83 TNA, BN 89/204, CSO, Committee for Statistics on Social Policy, *Study of the Better use of Administrative Data for Statistical Purposes*, 6 Apr. 1973, 2.
84 Ibid., 3.
85 Ibid., 4.
86 TNA, CAB 139/782, Moser, *Joint Approach to Social Policy: Statistical Implications, Briefing Notes for CSO Meeting*, 11th July 1975, 1.
87 Linda Challis, Susan Fuller, Melanie Henwood, Rudolf Klein, William Plowden, Adiran Webb, Peter Whittingham and Gerald Wistow, *Joint Approaches to Social Policy: Rationality and Practice* (Cambridge: CUP, 1988), 98–101.
88 TNA, BN 89/204, *PAR on Social Statistics*, 12 May. 1973, 2.
89 Anthony Wedgewood Benn, *The Regeneration of Britain* (London: Victor Gollancz, 1965), 15.
90 C. Anthony R. Crosland, *The Future of Socialism* (London: Jonathan Cape, 1964), 96.
91 Harold Wilson, "Statistics and Decision-Making in Government – Bradshaw Revisited," *Journal of the Royal Statistical Society. Series A* 136 (1973): 1–20, 7.
92 TNA, CAB 139/742, *Draft Speech for the RSS*, 10 September, 1969, 12.
93 TNA, BN 89/204, *PAR on Social Statistics*, 14 May 1973, 4–5.
94 TNA, BN 89/202, CSO, Committee on Statistics for Social Policy, *Meeting*, 9 Nov. 1972, 4 and 5.
95 TNS, RG 22/54, Redfern, *Personal Numbers and Population Registers: Background Note*, 21 Feb 1973, 1.
96 TNA, CAB 139/741, *Visit to EEC Statistical Office at Brussels on 28 and 29 March*, 9 Apr. 1962, 1 and 2.
97 Peter Hall, "Computer Privacy," *New Society*, 31 July 1969, 163.
98 TNA, RG 28/306, *Proposals for a Population Register System in Great Britain*, 3 Oct. 1967, 1.
99 TNA, RG 28/306, MacLeod to Rooke-Matthews, 3 Oct. 1967, 1.
100 TNA, PREM 16/1884, Moser, *Review of the GSS*, 4 Aug. 1978, 67.

101 TNA, CAB 108/417, CSO, *Report on Data Management Requirements*, 29 Nov. 1978, 11: and, TNA, BN 89/204, PAR on Social Statistics, 14 May 1973, 7.
102 TNA, CAB 108/417, CSO, *Report on Data Management Requirements*, 29 Nov. 1978, 13.
103 TNA, CAB 108/782, CSO, Working Party on GHS, *The Working of the GHS*, 26 Sept. 1974, 1.
104 *Civil Registration: Vital Change, Birth, Marriage and Death Registration in the Twenty-First Century*, Cm. 5355 (Jan. 2000), 25.
105 Cabinet Office, *Identity Fraud: A Study*, July 2002, 36–43.
106 TNA, HO 411/59, S. V. Cunliffe, *Brief for the M o S from the Statistical Department*, 19 Nov. 1975, 1–3.
107 TNA, HO 411/59, Home Affairs Committee, *PAR on the Subject of Social Statistics for Interdepartmental Policies*, 18 Nov. 1975, 1.
108 TNA, CAB 108/417, CSO, *Report on Data Management Requirements*, 29 Nov. 1978, 8.
109 Ibid.

4 Government 1979–97

Developing and Repurposing the Data State

On 3 May 1979, Margaret Thatcher became prime minister ushering in eighteen years of Conservative Party government. When she took office, she inherited a Central Statistical Office (CSO) that had more than doubled in size since 1964, while across the Government Statistical Service (GSS), the number of staff involved in gathering social data had increased from sixty-three, in 1965, to one hundred and forty-nine in 1979.[1] John Boreham, the chief statistical officer (Moser's successor, he held this post 1978–85), recognised that Thatcher's arrival presented the GSS with both threats and opportunities. On the one hand, the threats came from the government's aiming to "educate the public in 'economic realities'" a policy that it immediately applied to the state itself by driving through a series of programmes to cut the size of the government machine.[2] However, on the other hand, this presented opportunities because this desire to shrink the state went hand-in-hand with a desire to strengthen it. This may appear paradoxical but, as Foucault remarked, a government "which respects the specificity of the economy, will be a government that manages civil society, the nation, society, the social."[3] This is so because the free-market relationships pursued by the Thatcher governments were not natural phenomena. They needed to be installed and maintained, meaning that society and the population had to be reformulated to fit these requirements. Thus, the educational function that Boreham saw the government would pursue after 1979 was, by its nature, a biopolitical intervention designed to reshape the public into actors in a free market and this would, Boreham realised, require population data.

The way these apparently countervailing pressures unfolded is analysed in the first section of this chapter. This indicates how the government's need for population data, when paired with its desire to cut expenditure, led to an increasing centralisation of the GSS. This process unfolded in three stages. First, through the Rayner Review of 1980, second, the Pickford Report of 1988 and third, in the events that followed the arrival of Bill McLennan as chief statistical officer in 1992 and led to the formation of the Office for National Statistics (ONS) in 1996. It is demonstrated here that this institutional centralisation both reflected and reinforced an increased focus on the linkage of government datasets to provide the information government needed.

DOI: 10.4324/9781003252504-5

This centralisation represented an intensification of processes that had started before 1979; however, the following two sections of this chapter look at ways in which these governments sought to use population data to pursue policies that were different to their predecessors. When, in 1980, John Boreham drew up an annotated list of the policies that the Thatcher governments would need more data to enact, he included, in the category of social statistics, immigration and race relations along with health and welfare with this latter category encompassing the "encouragement of voluntary services and self-help, the concentration of State Welfare Services on those in real need."[4] The impact of these two sets of policies on the government's data-gathering operations is examined in the rest of this chapter with the second section focusing on immigration, while the third looks at fraud in the social-security system.

1

Soon after taking office, Thatcher appointed Sir Derek Rayner (he had been a senior manager at Marks and Spencer) as her special adviser on efficiency. He conducted reviews of the workings of Whitehall that included the government's data-gathering arm, the GSS. These reviews centred on challenging the "generalised arguments for preserving the status quo," which, he argued, had gone unquestioned for too long and had allowed a "coral reef effect" to develop, where data built up over time simply because there was already some data in place.[5] Rayner did not argue that government could work without any data at all. His report on the CSO stated that, for better or worse, "a highly numerate approach to the analysis of macro-economic policy issues and options has become part of central economic decision making."[6] But the phrase that resonates through all Rayner's reports is "value for money."[7] Data was too often seen as being good in its own right, without any regard being paid to the cost of gathering or storing it.[8] This in its turn raised the question of why the coral reefs of data had been allowed to develop at all. In addressing this, Rayner identified three problems that were inherent in the existing system.

The first of these was that, when government policy changed, data that was no longer of central importance was still collected and processed. For example, Rayner cited the work the CSO did on synthesising data from *Social Trends* for the Joint Approach to Social Policy (JASP). This policy had ceased and, he insisted: "it is right to expect the CSO to respond to those changes."[9] To encourage this change, Rayner recommended that officials should present an annual inventory of the data they collected. Through this, they should inform ministers of all the statistics collected using statutory powers, the timeliness of results drawn from these, when each dataset had last received political approval, a summary of all complaints received from the public for each corpus of data and when the next oversight meeting would be held.[10]

The second issue presented by the workings of the GSS was that, according to Rayner, data was held for purposes that far exceeded what the government needed. Data, he insisted, was not "a free resource" and it was not necessarily the government's role to provide it, as a public service, for anyone outside government who might want to use it. Rather, "it should be collected primarily because government needs it for its own business."[11] Rayner's report accepted that publishing data was "an important aspect of informing the public in a democracy and of maintaining public credibility in the government's figures." However, he also insisted that: "none of these considerations requires the government to bear all the costs of dissemination itself."[12]

The third issue identified by Rayner concerned the quality of service provided by the GSS. On this point, he did not argue that the service was too lackadaisical, in fact, the reverse was the case. While reviewing the service in the Home Office, it was suggested that: "there is a tendency for too much professional effort to be devoted to improving accuracy unnecessarily to the neglect of other work."[13] Nor was this zealous attention to professional standards confined to the Home Office, indeed the main culprit identified by Rayner was the Office for Population Censuses and Surveys (OPCS). This office was, Rayner had been told by its customer departments, "too research minded, too perfectionist." It was too keen to offer a "Rolls Royce" service rather than the "budget" offering many would have gladly accepted.[14] The CSO was obliged to agree with Rayner's views when it summarised them for the cabinet secretary, Sir Robert Armstrong. Here it noted, somewhat defensively, that there was nothing wrong with this professionalism as such, but accepted that its members needed to understand the difference between work that was essential and that which was merely desirable. Commensurate with this, it was necessary that someone in the chain of command should have the political-bureaucratic heft to cut off the flow at an appropriate point.[15]

Rayner thus identified the existence of an enclosure of statisticians with their own professional ethos as one of the factors increasing the numbers of data gatherers at work in the GSS and the scope of the work they undertook. But more than this, he identified the nature of their work itself as a driving force. This was most clearly seen in the review of the Home Office's statistical service, which noted that there was not any objective point at which work on statistics might be said to have ended. Data gathering in this sense was not like putting prisoners into cells because whilst when every prisoner was locked up, the job could be objectively described as complete, data could always be refined and new sets could always be sought. Thus, "inquiry and discussion have therefore failed to identify uniform objective criteria for determining where present operations should cease and fresh ones should be inaugurated."[16] Consequently, given that the whole process of data gathering was driven by "aspirational and qualitative" criteria, there was nothing inherently wrong with drawing a line at any particular point and preventing further collecting of data.[17] Rayner speculated that this might not have already happened because managers were awed by the technical nature of the subject, cowed by the professional mystic of statisticians or thwarted by the strength of their professional bulwarks. But regardless of the reason his report made it clear that "managerial values must be asserted."[18]

With this reasoning, Rayner had put his finger on the chimerical nature of the population-data gathering project. But Rayner was not actually trying to halt this pursuit. Indeed, as noted above, he began his review by recognising that data was necessary to government. Rather, his endeavours were directed at trying to cut the overall costs of these processes. In fact, from his managerialist perspective, getting more data would have been a desirable outcome as long as it were obtained more cheaply. Rayner seemed to anticipate that this energised assertion of managerialism would involve engaging in some considerable political tussles, and for sure there was, as will be seen, opposition to his proposals. However, his suggestions for how the GSS could do more work with reduced resources gave voice to two ideas that had long been current with the ranks of government data gatherers.

The first of these was that the GSS needed to use the data it held to a much greater extent than it was already doing. The GSS invested a lot of time and money on research to gather data and then presenting it in as close to a perfect state of accuracy as it could. Yet Whitehall departments already held a lot of information that was "not being used to best advantage."[19] Rayner's main recommendation here was that the GSS should increase its rate of computerisation and that it should review the work that had been done already in this regard since many government systems were not compatible and so could not share information.[20] What Rayner was driving at was that the government needed to share its data across departments. In other words, he gave a fillip to exactly the sort of data-linkage programmes that the government's data gatherers had been pursuing since the mid-1960s, and which people such as Acheson had been advocating for even longer. In fact even while Rayner was at work, before his reports came out, Boreham wrote to GSS staff setting out a strategy by which the service could survive the coming austerity axe with its professional pride intact. He was clear that increased data linkage was the key to this, arguing that social statistics "will, in future, be based on less data collection, with greater exploitation and interpretation of what is already available, including administrative sources."[21] Research projects would still be necessary, he maintained, to assess any emerging social problems, but for everything else, the first tool used should be "the linking of data within departments (e.g. within the DHSS [Department of Health and Social Security]) and across departments."[22]

The second way in which Rayner echoed and emphasised thinking within the GSS was that the drive to cut costs and increase data usage through increased linkage also became a means to further chip away at the codes of confidentiality that stood in the way of realising these aims. Rayner did not mention these codes in his reports but it is noticeable that the CSO brought them to the fore when it prepared an advance summary of points to discuss with the cabinet secretary, Sir Robert Armstrong. Thus, it highlighted how "protecting the confidentiality of data may conflict with using them efficiently." Showing very clearly how far the CSO, and indeed government more generally, had travelled since the days before Wilson became prime minister, this memorandum referred to the ways in which "*excessive* protection of confidentiality can inhibit legitimate research" and how, by doing do so it could prevent government data flowing into "the universities and institutes where it belongs."[23]

The boost Rayner gave to these tendencies within the government's data-gathering drive also increased its tendency towards centralisation because data linkage necessitated standardisation (usually through the use of some form of common numbering system) and this required a central index to be held. However, Rayner also promoted centralisation of the disparate structures of the GSS as an institution in more overt ways through his pursuit of economies of scale. One of these concerned the duplication of functions between Scotland and the rest of the United Kingdom. This led to difficulties in aligning the census findings from the OPCS with those from the General Register Office (Scotland) (GRO). Rayner accepted that some difference between the two would need to remain but could not understand why the gulf between them had ever been allowed to become wide enough to consume "a great deal of money" in the first place. But given that this was the situation he faced, he stressed the need for greatly enhanced cooperation between the two bodies.[24]

However, and more importantly, Rayner's clear avowal of the need for an assertion of managerial values meant that there needed to be someone with the necessary institutional standing to lead this charge. This person was to be the chief statistical officer whose role was changed as a result of Rayner's work. Rayner recommended that the post holder should become the person to whom ministers "can look to for advice on efficiency in statistics."[25] As was seen above, the reports voiced the sentiment that the non-specialist leaders of a department might be overawed by their departmental statisticians, were this to occur they could, from this point on, go over the heads of these in-house experts and seek the advice of the chief statistical officer who would have the government-wide reach derived from a new remit as head of profession.[26] A survey of the Rayner processes, conducted by two researchers from the University of Southampton, could thus conclude that the review process marked a shift of power away from departments to the CSO in general and to its head in particular.[27]

None of this posed an existential threat to either government data or to the GSS that gathered it. In fact, it presented opportunities to further policies, such as increased interdepartmental data linkage, which many within the GSS had been pursuing for years. However, the Rayner reviews did lead to considerable reshaping of the organisation through cutbacks and centralisation and there were losers in all this. As well as being "too perfectionist," the OPCS had been lambasted for being "too research-minded" and much of Rayner's pruning was directed at these activities. After asking, "why government needs its own social survey when there is a well developed private sector service," which was obviously a wholly rhetorical question, he went on to recommend cuts in the sample sizes used by surveys such as the GHS, along with the merger of the Family Expenditure Survey and the National Food Survey (see Chapter 3).[28] Boreham, who accepted Rayner's reports and therefore implemented the recommended 25% cut across the GSS, argued here that the OPCS's surveys could not be justified on the grounds of providing material to assess and change government policy. Rather, he maintained that it was the GSS's place to follow "conventional wisdom" as it had been redefined by the Thatcher government.[29] Research would no longer exist within government to fulfil David Donnison's aims of holding policy up to the light of evidence rather, under the new wisdom, its role would be to provide the data government needed.

Though this left surveys such as the GHS in place, it altered the way they were run. Moreover, the whole thrust of the Rayner Review impinged heavily on the enclosures that professional data gatherers had built up over the previous fifteen years, and predictably there was a backlash. Indeed eleven years later, the Royal Statistical Society (RSS) was still fuming about Rayner's impact in reducing the international standing of the British government's statistics through his "harmful" impact on the quality of analysis. In the short term, this came to a head at a meeting of the RSS held in June 1981.[30] At this meeting, Boreham made a statement, largely outlining the Rayner proposals and offering a rationale for them. For the remainder of the meeting, he sat and listened to a "very forceful" discussion, conducted in an atmosphere described as "electric" without a chance to respond. This discussion was led by Peter Townsend, Claus Moser and Alison Macfarlane (she was a medical statistician at the National Perinatal Epidemiology Unit in Oxford) and was stringent in its criticism of Rayner and therefore, by clear implication, of Boreham too. The report's view of "conventional wisdom" was described as emanating from "the 'don't confuse us with facts' school" with criticism being directed at two of its main features.[31]

The first of these was that the government's cuts to the GSS amounted "to a restriction of democracy."[32] In making this point, Townsend, unlike Boreham, concurred with Donnison and argued that in a democracy, it was the role of data, and those who held it, to hold government policies up to the light of solid evidence so that they could be seen for what they were and assessed accordingly. The second element of Rayner's report discussed at the meeting was the impact of its proposals on the ability of the GSS to act professionally. Moser had a lot to say on this point. In his opening address, Boreham had stated how the Rayner proposals would not prevent the GSS from acting in accordance with the general principles established by his predecessor, Moser, the most important of which was "to ensure that statistical integrity always wins the day."[33] However, Moser did not agree at all that Rayner's proposals were compatible with this, and to build his case, he brought forward three main points. First, he stated that the cuts would limit the GSS to gathering data for what government needed now, which could restrict the evidential basis of policy in the future should needs change. In Moser's words "one would think that judging the need for statistics was like judging the need for paper clips."[34] Second, he argued that cutting the extent to which data was checked and verified was a serious mistake that could lead to erroneous information being presented as accurate. Third, he expressed concern about the position of the OPCS. It was, he said, "sheer madness to cut the Social Survey Division which was probably the best survey organization in the country."[35] Rayner proposed removing all ad hoc surveying from the OPCS and this would, Moser suggested, weaken the organisation's infrastructure and thus impact the whole of the GSS not just the OPCS itself.

Boreham did not reply to criticism at the meeting and a subsequent written statement was an anodyne defence of Rayner. However, Boreham had been Moser's deputy from 1973 to 1978, and as such was presumably aware of four relevant points. The first of these was Moser's view of the technical inadequacies of the OPCS expressed in reply to Redfern's 1976 report (see Chapter 2). The second was that Moser had argued for an increased centralisation of data operations. Third, he must similarly have realised that Moser, like Townsend (and Boreham himself), wanted to break down the traditions of British data confidentiality without consulting the people and fourth, that everyone at the meeting (it was attended by 200 statisticians) probably wanted increased data linkage. In other words, while Rayner's report clearly unleashed an attack on the precise form and boundaries of the GSS's professional enclosure, and while it undoubtedly had an impact on the organisation's ability to produce accurate figures, it remains the case that in many respects it channelled pre-existing streams within the GSS so that they became stronger. As research projects such as the GHS faded in importance, data-linkage systems rose to fill the needs of government and centralisation became increasingly the order of the day. In this way, Rayner's impact was not to halt government data harvesting, rather the report put practitioners in a position where they were obliged to trade off some aspects of their processes against others.

The type of data-linkage systems that these governments brought forward as a result of these trade-offs are examined in subsequent chapters and would all add to the centralisation of data gathering and holding. But for now, it is necessary to show how, after Rayner, the pressure to increase the centralisation of the GSS as an institution continued. The next significant step in this process was the scrutiny of government economic statistics drawn up by Stephen Pickford, the head of the Economic Briefing Team in the Treasury, in September 1988. This scrutiny had been set up in response to charges raised by Conservative ministers that their handling of the economy was being misled by poor-quality statistics. As such, it focused on economic matters. Here, it both established that there was a problem with the quality of statistics and examined the technical reasons for this. However, in the course of their investigations, Pickford and his team came to the conclusion that these technical shortcomings were exacerbated, if not caused by, the nature of the current system. They discovered that effort that could have been applied to producing accurate data was instead diverted into managing the system itself and that this was needed because of the decentralised shape of the GSS. The CSO had the responsibility for compiling accounts, but to do, so it relied on data collected by GSS staff working in other departments. These other departments habitually brushed aside the CSO's attempts to streamline these connections as they conflicted with departmental priorities. A system of interdepartmental committees existed with the brief of sorting out such matters, but these bodies were ineffective talking shops. In other words, there was no one person or office actually in charge of the system with the authority to impose standards and this had allowed the whole body to drift.

This state of affairs, which seemed to come as a surprise to Pickford, led to the conclusion that "the CSO's current internal organisation is not appropriate for a co-ordinated approach to compiling national accounts."[36] As a result of this diagnosis, he produced a variety of suggested reforms that ran up to and included making "the CSO an agency of the Treasury while taking account of other Departments' needs and priorities."[37] These proposals struck a chord with the higher echelons of the civil service. Sir Robin Butler, the head of the service, called a meeting to discuss them and a briefing paper issued in advance of this described Pickford's proposals, "as a whole," as: "representing a step change" in the way economic data was used.[38] There was awareness that a lot of the proposals were "contentious," and that discussion could therefore easily be side tracked into dealing with "quibbles over individual recommendations." However, the organisers were determined to prevent matters from degenerating into a process that fixed short-term issues while leaving the structural problems of the system unreformed.[39] As part of this drive to implement the whole Pickford package, the idea that direct control by the Treasury might compromise the integrity of statistical output, or that it might be seen to be doing so, was simply dismissed.[40]

However, the chief statistical officer, Jack Hibbert (Boreham's successor, he held this post from 1985 to 1992), did not agree. He argued that significant aspects of his role necessitated his office being institutionally sited outside of any one ministry with its current location, in the Cabinet Office, providing an ideal base. These aspects of his office included, dealing in an even-handed way with all departments across Whitehall, liaising with international bodies, maintaining contacts and relationships with the wider statistical community and "running through all these ... is the need for oversight ... of the professional staff in the GSS."[41] Moreover, Hibbert leaked news of Pickford's proposals to a wider group of permanent secretaries than had been privy to the plans and this galvanised some of them to fall in behind his resistance to moving the CSO to the Treasury.[42] In addition to this support, the RSS held a well-attended meeting of its members to discuss the issue of public confidence in the integrity and validity of official statistics.

Hibbert opened this meeting with a short paper that did no more than outline some issues pertaining to the topic, but this was all that was required to open the floodgates to a torrent of support for his opposition to a Treasury takeover of the CSO. There was widespread feeling at the meeting that if, as government maintained, the statistics it received were not up to scratch then, in Moser's words "government has only itself to blame" because the GSS had been eviscerated by Rayner.[43] Rather than relocating the CSO, described as "a serious error," many speakers agreed with Moser that what the country needed was a national statistical council.[44] This would bring together key interests from both inside and outside government, a forum for the exchange of ideas and concerns and would support the GSS against political encroachments on its professional enclosure.[45]

In the end, Pickford's proposals were not fully implemented. However, the Business Statistics Office was moved from the Department of Trade and Industry to the CSO. In essence, this meant that the CSO had become an office that supplied economic data to its main customer, the Treasury, although it still retained oversight of the rest of the GSS. This was something of a halfway house between the status quo that had existed before Pickford, and what Pickford had sought to implement. However, subsequently, in November 1991, the CSO was detached from the Cabinet Office to become an executive agency. Under these new arrangements, the chancellor worked with the director of the CSO to determine its resources, while the director received advice from a management board and an advisory committee representing the views of the CSO's main customer departments. However, the director alone had control over professional standards and the selection of appropriate methodologies for data collection.[46] This was the position when Bill McLennan became the CSO in 1992.

On taking up the post, McLennan wrote that he wanted to make changes "for the longer term." He described this in a letter to the chancellor, Ken Clarke, as

> doing something about the poor state of government statistics in the UK and not ... just trying to influence changes in the short term. I am trying to leave behind something which will continue to grow when I return to Australia. This is why structural changes are so important.[47]

These structural changes would all centre on increasing the centralisation of the institutions at the heart of the data state.

In January 1993, soon after he took office, McLennan received a lengthy memorandum on *Further Centralisation of Statistics in the UK*. This made the point that, given the history of Britain's decentralised data gathering structures, simply making the point that centralisation was, in itself, a good thing would not win any political arguments. Thus, if he wanted to make the case for centralisation, McLennan was going to have to find a systemic problem that he could claim this would resolve. The author gave an approving outline of the work done by Moser and Pickford, and McLennan would always argue that his proposals followed in their footsteps. In June 1993, he wrote to Sir Robin Butler noting that he approved strongly of Pickford's work in furthering centralisation and that "these changes have been extremely successful."[48] Similarly, in the same letter, he noted that within the system created in Britain in the late 1960s, Moser had always taken "deliberate steps to achieve a degree of centralisation and central management."[49] The writer of the January 1993 memorandum listed a variety of the government's data operations that McLennan might seek to absorb into the CSO. This included work currently done by the OPCS, the employment statistics, construction statistics or major collections of social data. Alternatively, it was suggested that the new chief statistical officer could seek to tie the whole GSS more closely together through a series of codes of conduct, or other administrative reforms. But whichever course of action was pursued, it was reiterated that simply stating the benefits of centralisation would not win the argument to get it. In other words, if McLennan really wanted to do something about what he saw as the poor state of British government statistics, he would need to outline the root of these shortcomings and show how centralisation was the solution.

When he wrote to Ken Clarke, McLennan gave the chancellor "some frank background" highlighting what he saw as the real issue. This was that the "OPCS is not very well managed ... and does not influence government policies."[50] Moreover, McLennan knew that the OPCS was likely to become an agency in the near future, which would mean that he would find it harder to control. This meant that he needed to do something about the situation quite quickly and this timetabling was made more urgent by the fact that both the OPCS director, Peter Wormald, and his deputy, Eric Thompson, were due to retire in the next two years. McLennan thus had a window of opportunity and he was determined not to let it close on him.

However, none of the shortcomings he saw in the operations of the OPCS necessarily meant that the CSO needed to take over the whole of that organisation, which was the plan that McLennan revealed to Clarke. Opponents of this move always argued that the same benefits could be achieved by introducing mechanisms to build closer coordination between the two bodies. Thus, in writing to Clarke, the secretary of state for health, Virginia Bottomley, whose office had ministerial oversight for the OPCS, noted that, while she strongly supported improving the efficiency and quality of the government statistical machine, "I think it is possible to further that simply by better collaboration between CSO and OPCS."[51] Similarly, at a meeting between McLennan and Wormald, held under the auspices of Sir Robin Butler in June 1994, Wormald argued that the improvements in the standards of data that everyone wanted did not necessarily depend on a merger of his organisation with the CSO. Moreover, he also produced two more reasons why the merger could not happen. The first was that the registrar general could not, without changes in primary legislation, become subject to any other ministerial authority. Second, that the medical profession would be extremely chary about having medical data, held by the OPCS, placed under the control of the prime minister, which is where this information would be held, were the OPCS to be taken over by the CSO from its Cabinet Office base.[52]

However, McLennan insisted that while increased coordination between the two offices could provide opportunities for improvements in government data, this did not mean that it would actually deliver these advances.[53] Here, he based his conclusion on his understanding of the recent history of the CSO and the GSS more generally. He argued that the system established by Moser worked well enough when Moser was in charge. At that time, Moser had a strong political relationship with the prime minister, which gave him the influence to win Whitehall battles and to keep all the components of the system not only working, but working harmoniously. However, in McLennan's view, after Moser left the CSO, things did not operate "so smoothly and parts of the system he set up appear to have slowly disintegrated," to the extent that the only thing that held the system in any sort of alignment was the professional pride of statisticians.[54] What McLennan was driving at was that Moser's system was just that: it was Moser's, and it depended on a forceful personality, political clout and goodwill to keep working. McLennan had no doubt that he could deliver such a system, but he also argued that if he did, the example of his revered predecessor would indicate that such a legacy would quickly come to nothing when he (McLennan) returned to Australia. What McLennan wanted was unified structures to institutionalise change.[55]

If Moser had had the ear of the prime minister, McLennan was not adverse to using personal connections to further political ends. He went on golfing weekends with Sir Robin Butler and took the Ken Clarke out "for a beer (or two)" to have the opportunity to persuade the chancellor to get the prime minister's backing to "fix" the "OPCS problem" by supporting the merger. Whether this networking was the reason must remain unclear, nevertheless, it is the case that when Clarke wrote to Bottomley, he repeated verbatim the arguments made to him by McLennan and told her that the plan had the backing of the prime minister.[56] At this point, Bottomley's objections, such as they had been, ceased. Arguably the most serious thing she had done was to give ministerial voice to the hostility to the merger raised from within the enclosure of the medical profession concerning the sanctity of patient confidentiality. However, McLennan and his allies were able to circumvent this issue and the subsequent creation of the Office for National Statistics, in 1996, resolved all these complaints. This new organisation would become McLennan's legacy and his structural remedy for the paucity of government statistics in the United Kingdom.

However, this leaves a big question unanswered, namely what exactly was the problem that required such a drastic institutional realignment? An increase in centralised control was not something McLennan inaugurated simply for its own sake, and neither was it pursued to resolve professional shortcomings in the performance of statistical units along Whitehall. Rather the whole merger was fundamentally designed to further data linkage within government. Thus, McLennan's insight was that the structures of the GSS existed in a reciprocal relationship to the data the system held. Wanting to coordinate the data therefore necessitated centralising the structures rather than tinkering with their patterns of communication.[57]

In an article published in the Journal of the RSS, McLennan quoted approvingly an academic who had, with distinct echoes of the comments of Acheson, Titmuss and Morris from the early 1960s, described the output of the GSS as fragmented and inconsistent. There was, he said "no feeling of a statistical profile/compendium of Britain coming through them. Links between national statistics coming from different sources is tenuous, and links with the continuous surveys designed to supplement them, often non-existent."[58] In his frank letter to Clarke, McLennan noted that it was also his opinion that there was a growing need to address the problem of "the mass of information in the government statistical service, mostly coming from administrative sources, which is very under-utilised, and not related across sources."[59] Data linkage was, as has been shown in Chapter 3, a feature of the continental systems that the British government's data gatherers often aspired to emulate and, in this regard, McLennan was no different from others in the GSS. Indeed, he wrote to Butler that his proposed merger of the CSO and the OPCS would create opportunities for the British data system to make use of all the data currently held by the OPCS, adding that states in the EC in general, and in Scandinavia in particular, had already made "significant" moves in this direction, which "must be addressed in a serious way in the UK."[60]

McLennan's strategic point here was that achieving this would require much more than simply improving the coordination between two offices that remained separate. What was needed was a full merger that would bring all the data together in one organisation where all these linkages could be institutionally welded into a closed loop. Not only would this office hold all the data, it would also bring together all the skilled people required by such a project who would be able to cooperate on its analysis and use, free from the pressures and exclusions of Whitehall turf wars. The merger would thus, "make it easier and feasible to bring together the massive amount of data currently existing in government, to relate these in a meaningful way, and to make them available across government."[61] Here it is important to note that it was this desire to link data that led to the centralisation. Thus, these events highlight how, as Foucault expressed it, the pursuit of population data would always be the "discourse of a centred, and centralising power."[62] Given the rhetorical focus of these governments on cutting the size of the state, this may seem paradoxical. But while taking an axe to budgets, these governments simultaneously placed increased demands on their data gatherers. It was this demand, for more data to be provided for less outlay that predisposed McLennan and others to heighten the longstanding desire to link administrative datasets rather than conduct research to gather more. The increasing demand for data in this political climate formed McLennan's vision for the GSS and his desire to drive through a "paradigm shift in the responsibilities of official statisticians."[63] The rest of this chapter looks at two sets of policies that came to define these governments, both of which required increased flows of increasingly centralised population data.

2

On 20 April 1968, Enoch Powell made his "Rivers of Blood" speech in which he excoriated the government's immigration policy. This led to popular protests in favour of his positions that mobilised such support that it has been argued that his intervention was the reason for the Conservatives winning a surprise victory in the 1970 election.[64] Regardless of this argument, it remains a fact that from this point on, immigration became a high-profile political issue and, under the Thatcher/Major governments, it not only remained on the political agenda, it increased in importance. It was also used to justify further population-data gathering to the public and became a priority for the development of government data systems. For example, in 1982, when the Home Office was beginning to computerise its systems, it prioritised the installation of the new technology in its Immigration and Nationality Directorate.[65]

However, the most important example of how the strident moral panic around immigration spread across British politics to skew the government's data gathering can be seen in attempts, initiated under the previous Labour government, to introduce questions on ethnicity and nationality into the 1981 census. This followed Powell's allegations that the size of Britain's black population was larger than that given by the official figures: in other words that the government's data was wrong.[66] Thus, the 1978 white paper on the 1981 census stated that there was a need for authoritative data on Britain's main ethnic minority populations and that the registrar general would use the census to obtain this. A trial had already been conducted of possible questions to harvest this information. But on the eve of the 1979 general election, in an atmosphere poisoned by Thatcher's remarks about Britain being "swamped" by alien cultures, the registrar general's office conducted more tests in the London borough of Harringay where the whole enterprise was derailed by popular local opposition.

In defending these trials before a sub-committee of the House of Commons Home Affairs Select Committee in 1982, Roger Thatcher, the registrar general, gave his support to the policy of enquiring into the ethnic make-up of the population stating that this was necessary to provide the data that alone could ensure that community relations progressed on a solid basis of fact. However, he was aware of the practical problems posed in getting this information. First, simply asking people where they had been born would only get the required information for the first generation of immigrants since, as he put it in referring to people of West Indian origin, "they did not want to describe their children, who were born in this country, in this way. Instead, they left the question blank or described their children as 'British.'"[67] His point was that, as more children were born in Britain to people who had come to Britain from the Caribbean, the overall black population could increase, but in this manner still not be recorded as such in the official data. Second, the term "race" needed to be used. The word "ethnic" was too confusing because it was not widely understood, and the word "origin," used in the phrase "racial origin," was equally confusing as black people from the West Indies might describe their racial origins as being African. Third, there was the point that he was not attempting to assess the level of all migration into Britain, but that of people who were not white, therefore, since some West Indians were white, simply asking people where they had been born, or even where their parents had originated from, would not achieve the desired outcome.

The tests were carried out by distributing census forms, with half containing the new questions on race, to 56,000 households along with in-person follow-up interviews conducted with 2,000 people. Response rates to the forms varied from 56% for Asian households to 34% for West Indian households. Moreover, the follow-up interviews revealed that only 14% of the West Indian and 34% of the Asian households that had completed the question on race had done so correctly (Thatcher did not specify how an answer was assessed as being correct and no one on the committee asked for this information). These interviews also showed that even among those who had answered the question on their parent's country of birth, 37% (of both Asian and West Indian households) objected to this on principle.[68] It is thus clear that there was strong opposition to this data-gathering exercise and the registrar felt that this was so serious that it "might have upset the rest of the census."[69] Accordingly, on 20 March 1980, Patrick Jenkin, the secretary of state for social services, announced that the 1981 census would not contain any questions on ethnic origin, parents' place of birth, nationality or year of entry into the United Kingdom. These had been dropped, the OPCS announced, because any census needed to "be broadly acceptable to members of the public," criteria which this attempt had clearly failed to meet.[70]

In reacting to the evidence given by Roger Thatcher and his staff, the chair of the Home Affairs Select Committee made reference to "got up opposition" and to rumours spread "by evilly motivated extremist political groups" "who seek to exploit those questions for their own political ends."[71] This laid the blame for the non-participation rates in Harringay at the door of those who had composed and distributed 25,000 copies of a leaflet that claimed the census trial was linked to a proposed change in the United Kingdom's nationality laws that:

> would make nationality dependent on your parents' nationality, not where you were born ... If we say now who is or is not of British descent, we may one day be asked to 'go home' if we were born here or not.[72]

From all of this reciprocal hostility, the social scientist Martin Bulmer concluded, in 1986, that: "a British census question on race cannot be considered apart from the political climate and circumstances under which the census is conducted."[73] By the late 1970s/early 1980s, this climate had become so toxic that even though census data could, in theory at least, be used to channel beneficial government services towards minority communities, there was no lobbying, by either social scientists, or by groups representing these communities, in favour of including these questions in the 1981 census. However, what did exist at the time was clear evidence that the reaction found on the streets of Harringay should have been anticipated.

Chapter 3 showed how one response from within the OPCS to the government's clamping down on the cost of data gathering was to revamp the electoral roll as a form of population register. The way this scheme would have worked, had it been implemented, will be examined in Chapter 6. But for now, it is necessary to point out that when this policy was first devised, in 1977, the OPCS conducted extensive research to test the public's reaction. This was conducted in three areas, Bradford, Leicester and Stratford upon Avon, the first two of which contained relatively large populations of people from ethnic minority communities. This research found that one of the issues that caused most people to baulk at the proposed change to the electoral register concerned the way it enquired about migration.[74] This insight into the public's reaction already existed inside the OPCS, headed by the registrar general, before the field trails were conducted in Harringay. Given this, it seems clear that there must have been considerable political momentum behind the suggestion that these questions ought to be included in the census. Moreover, even if this were not the case, the importance of immigration did not decrease in British politics after the 1981 census. In fact though the focus changed slightly, to centre on the issue of illegal immigration, this would become one of the main ways in which government in this period would seek to justify its data trawls of the population.

3

One of the other key ways in which these governments would seek to justify the increasing depth and breadth, their data stocks of the British population was through claims that these could be used to combat fraud, particularly fraud in the social-security system. An interest by government in this issue was not in itself a new thing. At the first meeting of the Ministerial Group on the JASP, in 1975, Labour ministers discussed the way in which "abuse of the social-security services by those who shirked work in order to claim unemployment benefit was assuming great psychological importance, and could be highly damaging to the government's credibility among its own supporters."[75] What is particularly noteworthy here is that in summing up the meeting, the prime minister, Harold Wilson, made no further mention of this issue. However, under Conservative prime ministers after 1979, this issue would not be pushed off the agenda in this way, far from it.

Conservative governments recognised the potential popular resonance of this issue and acted accordingly. Thus, Peter Lilley was the first secretary of state (he held this post from 1992 until the Conservatives lost power in 1997) to require the Department for Social Security (DSS) to supply regular indices of social-security fraud. He also published an anti-fraud strategy that pledged the department to pursue fraudsters and heightened the role of fraud detection across departmental activities. This action was echoed by calls from the House of Commons Social Security Committee for tough action against fraud.[76] The governments Lilley served used this issue to both win support for their data trawls and to deflect attention away from other issues. For example, when, in 1988, the government attempted to introduce ID cards (see Chapter 7), it became concerned that these would be seen as a tool of the EC being forced onto Britain and in order to redirect any such criticism, ministers suggested that the cards should be presented as tools to be used against fraudsters.[77] Powell had linked the issues of the misuse of the welfare and health systems with immigration in his "Rivers of Blood" speech. He saw the systems of social government as a reward to the British people for the sacrifices of the Second World War, and since newly arrived immigrant communities could not, by his reasoning, have made a contribution to these efforts, they were not entitled to use these systems.[78] The deploying of the rhetorical device of the fraudster by these governments relied on similar reasoning. Thus, in 1994, in the midst of another government drive to introduce ID cards, the chancellor of the Duchy of Lancashire, William Waldegrave, told his colleagues that the proposals could "prevent social security fraud and false claims of entitlement to National Health Service treatment,"[79] all of which would provide fertile ground for the subsequent hue and cry beaten up by the Blair government against what it branded "health tourists."[80]

After 1979, a variety of ministers pursued ID cards on the basis that these innovations would combat social-security fraud despite the fact that they knew these cards could not achieve this goal. In 1988, John Moore, the secretary of state for health and social services, wrote to Douglas Hurd, the home secretary, telling him that there were three reasons why ID cards would be of little benefit in fighting social-security fraud. The first of these was that most claims were not made in person but by post. Second, most claimants, who were required to do so, could already provide adequate proof of identity and third, most fraud was not committed by people claiming to be someone else, but rather by people misrepresenting their true circumstances.[81] Similarly, at a later attempt to introduce ID cards, the Home Affairs Committee was told that the cards could only hope to prevent a maximum of 5% of all social-security fraud.[82] As a result fraud was really combatted by other means that centred on data linkage.

Systems for linking records that were developed in this period will be examined in more detail in Chapter 6, but for now, it is necessary to follow one commentator, writing in 1986, and note that linkage between computerised datasets allowed for checks to be made to prevent fraud, and that because this was possible, and because fraud was assumed, these links and checks were made. Thus:

> the DHSS fraud squad often target their enquiries on unemployed people with specific sets of skills (e.g. motor mechanic, carpenter) and unmarried or single women. The assumption is that unemployed people with skills are earning on the 'black economy' or that women are living with or being supported by a man.[83]

The government was thus clear that data linkage provided the best method to curtail social-security fraud. Therefore, this formed the basis of the system introduced by Peter Lilley. He had inherited a social-security system that did not have a central register of claimants and payments, rather the system held its data on separate "data islands" with separate computer systems that could not be linked. In the conclusion of three scholars, the "DSS seems to have accepted a certain level of fraud as an inevitable if regrettable consequence of the complex and fragmentary nature of the social security system."[84] Nevertheless, the government also believed that the image of the fraudster had such a hold on the popular political imagination that it continued to deploy this rhetoric as a justification for many of its data-gathering schemes, even though these could not be justified by any real impact on social-security fraud.

However, the reality was that schemes that were centred on technology, such as the smart cards brought forward by the government in the 1990s (see Chapters 7 and 8), risked opening up a new front in government systems to be attacked by fraudsters. As the Ministerial Group on Card Technology recognised, in 1994, "new technology, brings with it new methods of fraud."[85] Thus, just as British governments were, from the mid-1960s onwards, locked into the pursuit of the chimera of a perfect set of population data, so the focus on fraud opened up its own never-ending campaign to protect the systems that were designed to protect the state from fraud.

Of course this only applied once the political decision had been taken to make fraud into a big political issue. This never happened under Wilson, but it did under Thatcher and Major. As was seen above, Foucault remarked that when a government leaves the economy alone, it turns its attention to civil society. Indeed, he notes that economic freedom, the heart of the neo-liberal component of Thatcherism, is not a natural phenomenon: one that simply needs to be located and unleashed, rather it is something that is constantly produced by active political endeavours. This production and reproduction involves the liberal state in arming itself with "a precise, continuous, clear and distinct knowledge of what is taking place in society, in the market and in economic circuits."[86] It is at this point that the full implication of the issue of fraud can be discerned.

Conservative governments may have been keen to publicly embed their hostility to social-security fraudsters in the language of fairness, responsibility and prudence, but it is important to see these frauds (insofar as they existed) for what they were: they were frauds against the state. To an extent it did not really matter why the government wanted to know, "in the round," the individuals that comprised the population. In Chapter 3, it was shown how Wilson's Labour government had wanted to do so "to bring about compassionate as well as efficient societies," whereas Thatcher wanted similar knowledge to oblige people to conform to the rigours of the market.[87] Thus, what really mattered was the epistemological approach to policy, since by keeping their identity and lifestyle private, or by being inadvertently beyond the state's gaze, such individuals were frustrating the purpose of a modern state: to know the population. In the period after 1964, the state became determined to know the population, and to do this, it needed to know the individuals that comprised it. Those who committed such frauds were thus attempting to resist the advance of this knowledge.

The advocates of increasing government data always argued that when people agreed to be identified, they also agreed to be included in the community. Redfern wrote of how, in the United States, the census was viewed as a "national ceremony" and endorsed the words of the director of Statistics Sweden on this point, thus: "a population census has a ritual aspect in the life of a nation … To be counted in the census is then a manifestation of citizenship, of being one in the nation."[88] The corollary of this was that those who resisted identification, those who wanted to pick and choose from the obligations and rights inherent in membership of a modern polity were positioning themselves as outsiders. Redfern, stated that these "à la carte" elements of the population ("perhaps a few percent") included those who evaded their taxes, those who refused to register to vote (see Chapter 6) and, tellingly for this chapter, illegal immigrants along with those who would not complete a census form.[89] Redfern concluded his citation of the Swedish statistician thus: "a nation that cannot carry out a census … is a nation in trouble in other ways as well."[90] It was this perception of a crisis, "a step towards anarchy" was Redfern's apocalyptic phrase, striking to the heart of the government's ability to know, and therefore to govern the population, that produced the extraordinary splenetic outbursts ("evilly motivated extremist political groups," above) by the chair of the Home Affairs Select Committee in the face of the Harringay protests.[91]

Redfern argued that the solution to events like the Harringay protest or to endemic social-security fraud, was to be found in gathering data, and if people would not willingly provide it, he could not see any reason why the government should not develop an all-encompassing system that would prevent anyone from opting out. He advocated the development of a full, Scandinavian-style population register, or the adoption of the next best thing: a system of common numbering for all government files. He explicitly argued that either of these systems would act as a modern, data-based panopticon by promoting socially desirable qualities ("fairness in sharing burdens and benefits according to the rules, including ensuring that those entitled to benefits do in fact get them") while simultaneously acting as a brake on crime: "fraud, impersonation and the creation of false identities."[92] These types of systems, he argued, thus "imposed" "discipline" on the population, or as Boreham noted of Thatcher's intentions, could re-educate the people in the behaviour demanded by a free market.[93]

Notes

1 TNA, PREM 9/2, *Review of the Government Statistical Services, Report on the Central Statistical Office*, Oct. 1980, 5; and TNA, PREM 9/1, Sir Derek Rayner, *Review of Government Statistical Services, Report to the Prime Minister*, Dec. 1980, 31.
2 TNA, CAB 164/1557, *A Strategy for the Government Statistical Service, A Note by the Head of the GSS*, 30 Jan. 1980, 8.
3 Michel Foucault, *The Birth of Biopolitics* (New York: Palgrave MacMillan, 2010), 296.
4 TNA, CAB 164/1557, *Strategy for the GSS*, 30 Jan. 1980, 10.
5 TNA, PREM 9/1, Rayner, *Review of GSS, Report to the PM*, Dec. 1980, 11.
6 TNA, PREM 9/2, *Review of the GSS, Report on the CSO*, Oct. 1980, 6.
7 See, for example, Ibid.
8 TNA, PREM 9/1, Rayner, *Review of GSS, Report to the PM*, Dec. 1980, 9.
9 TNA, PREM 9/2, *Review of the GSS, Report on the CSO*, Oct. 1980, 6.
10 TNA, PREM 9/1, Rayner, *Review of GSS, Report to the PM*, Dec. 1980, 13.
11 Ibid., 11 and 17.
12 Ibid., 17.
13 Ibid., 2.
14 TNA, PREM 9/1, Rayner, *Review of GSS, Report to the PM*, Dec. 1980, 29 and 34.
15 TNA, CAB 164/1557, Beesley to Armstrong, Oct. 1980, 2.
16 TNA, PREM 9/12, *Review of the Home Office Statistical Services*, July 1980, 18.
17 Ibid., 17 and 20.
18 TNA, PREM 9/1, Rayner, *Review of GSS, Report to the PM*, Dec. 1980, 20.
19 TNA, PREM 9/12, *Home Office Statistical Services*, July 1980, 19.
20 TNA, PREM 9/1, Rayner, *Review of GSS, Report to the PM*, Dec. 1980, 8.
21 TNA, CAB 164/1557, *Strategy for the GSS*, 30 Jan. 1980, 3.
22 Ibid., 12–13.
23 TNA, CAB 164/1557, Beesley to Armstrong, Oct. 1980, 3. Emphasis added.
24 TNA, PREM 9/9, *Report of Review of Statistical Services in General Register Office (Scotland)*, June 1980, 15.
25 TNA, PREM 9/1, Rayner, *Review of GSS, Report to the PM*, Dec. 1980, 17.
26 *Government Statistical Services*, Cmnd. 8236 (Apr. 1981), 3.

27 TNA, CAB 164/1723, Gerald Hoinville and Terence Michael Frederick Smith, (1982) "The Rayner Review of Government Statistical Services," *Journal of the Royal Statistical Society, Series A* 145 (1982): 204.
28 TNA, PREM 9/1, Rayner, *Review of GSS, Report to the PM*, Dec. 1980, 31 and 33.
29 Hoinville and Smith, (1982) "The Rayner Review," 206.
30 The Royal Statistical Society (1991) "Official Statistics: Counting with Confidence. The Report of a Working Party on Official Statistics," *The Journal of the Royal Statistical Society, Series A* 154 (1991): 26.
31 Hoinville and Smith, "The Rayner Review," 198.
32 Ibid., 197
33 Ibid., 196.
34 Ibid., 197.
35 Ibid., 199.
36 TNA, CAB 164/1910/1, *Scrutiny on Government Economic Statistics – Synopsis*, 7.
37 Ibid., 11.
38 TNA, CAB 164/1910/1, *Statistics: Meeting*, 9 Dec. 1988, 1.
39 Ibid., 1–2.
40 Ibid., 2.
41 TNA, CAB 164/1910/1, *Responsibilities of the Directors of the Central Statistical Office (CSO) and Head of the Government Statistical Services (GSS)*, 17 Jan. 1989, 1–3.
42 TNA, CAB 164/1910/1, Spencer to Sir Robin Butler, 1 Feb. 1989, 2.
43 Jack Hibbert, "Public Confidence in the Integrity and Validity of Official Statistics," *Journal of the Royal Statistical Society* 153 (1990): 130.
44 Ibid., 142.
45 Ibid., 130.
46 Bill McLennan, "You can Count on Us – With Confidence," *Journal of the Royal Statistical Society. Series A* 158 (1995): 472.
47 TNA, RG 50/36, McLennan to Clarke, 16 Dec. 1993, 4.
48 TNA, RG 50/35, McLennan to Butler, 9 June 1993, 4.
49 Ibid., 2.
50 TNA, RG 50/36, McLennan to Clarke, 16 Dec. 1993, 1.
51 TNA, RG 50/36, Bottomley to Clarke, 16 Nov. 1994, 1.
52 TNA, RG 50/35, *Note for the Record: CSO/OPCS Amalgamation*, 24 June 1993, 2–3.
53 TNA, RG 50/35, *CSO/OPCS Amalgamation, Paper by W. McLennan and P. J. Wormald*, n.d., ca. July 1993, 3.
54 TNA, RG 50/35, McLennan to Butler, 9 June 1993, p.3; and TNA, RG 50/35, *Note for the Record*, 24 June 1993, 1.
55 TNA, RG 50/35, *Meeting to Discuss CSO/OPCS Amalgamation*, 23 June 1993, 2.
56 TNA, RG 50/35, McLennan to Butler, 11 May 1993, 2; and, TNA, RG 50/36, McLennan to Chancellor, 1; and, TNA, RG 50/36, McLennan to Butler, 26 Sept. 1994, 1; and, TNA, RG 50/36, Clarke to Bottomley, 17 Oct. 1994, 1.
57 TNA, RG 50/35, GSS Committee on Social Statistics, *Meeting*, 15 Mar. 1993, 5.
58 McLennan, "You can Count on Us," 480.
59 TNA, RG 50/36, McLennan to Clarke, 16 Dec. 1993, 3.
60 TNA, RG 50/35, McLennan to Butler, 9 June 1993, 6.
61 TNA, RG 50/36, McLennan, *Possible Merger of OPCS and CSO*, 26 June 1994, 2.
62 Michel Foucault, *Society Must Be Defended* (London: Penguin, 2004), 61.
63 McLennan, "You can Count on Us," 475.
64 Nikolas Rose, *Governing the Soul: The Shaping of the Private Self* (London: Free Association Books, 1999), 197.
65 TNA, HO 524/1, IND Computer Strategy Committee, *Meeting, 12 July 1982*, and TNA, HO 524/1, IND Computer Strategy Committee, *Meeting*, 21 July 1982.

66 Martin Bulmer, "A Controversial Census Topic: Race and Ethnicity in the British Census," *Journal of Official Statistics* 2 (1986): 472.
67 House of Commons, Home Affairs Select Committee, Race Relations and Immigration Sub-Committee, Ethnic and Racial Questions in the Census, Minutes of Evidence, 22 Nov. 1982, *Memorandum Presented by the Registrar General for England and Wales*, 57.
68 Ibid., 58.
69 Ibid., 107.
70 House of Commons, Home Affairs Select Committee, Race Relations and Immigration Sub-Committee, Ethnic and Racial Questions in the Census, Minutes of Evidence, 22 Nov. 1982, *Appendix C, OPCS Monitor*, 2 Apr. 1980, 81.
71 Ibid., *Examination of Witnesses*, 322, 316, and 281, *Mr. Thatcher*, 109, 108 and 105.
72 Quoted in, Bulmer, "A Controversial Census," 474.
73 Ibid., 477.
74 OPCS Population Statistics Division, *Report of the Steering Committee to the Registrars General, Extending the Electoral Register – 1* (London: OPCS Occasional Paper, 20, 1981), 11.
75 TNA, T 227/4396, Ministerial Group on a Joint Approach to Social Policies, *First Meeting*, 13 May 1975, 3.
76 Perri, 6, Charles Raab and Christine Bellamy, "Joined-up Government and Privacy in the United Kingdom: Managing Tensions between Data Protection and Social Policy. Part 1," *Public Administration* 83 (2005): 398–399.
77 TNA, FCO 30/7576, Identity Cards, 31 Aug. 1988, 4.
78 See, Camilla Schofield, *Enoch Powell and the Making of Postcolonial Britain* (Cambridge: CUP, 2013).
79 TNA, CAB 130/1486, Ministerial Group on Card Technology, *Meeting*, 30 June 1994, 1.
80 Dept. of Health, *Proposals to Exclude Overseas Visitors from Eligibility to Free NHS Primary Medical Services*, May 2004.
81 TNA, LAB 109/88, Moore to Hurd, 14 Sept. 1988, 1.
82 Home Affairs Committee, *Fourth Report, Identity Cards, Vol. 1*, 26 June 1996, vii.
83 C. Pounder, "Police Computers and the Metropolitan Police," *Information Age* 18 (1986): 4.
84 6, Raab and Bellamy, "Joined-up Government," 398–399.
85 TNA, CAB 130/1486, Ministerial Group on Card Technology, *Meeting*, 24 May 1994, *Smartcard Report*, 10.
86 Foucault, *The Birth of Biopolitics*, 64 and 62.
87 TNA, CAB 139/742, *Draft Speech for the Royal Society of Statisticians Banquet on the Occasion of the 37th Session of the International Statistical Institute*, 10 Sept. 1969, 12.
88 Philip Redfern, "A Population Register or Identity Cards for 1992?," *Public Administration* 68 (1990): 506; and Philip Redfern, "Population Registers: Some Administrative and Statistical Pros and Cons," *Journal of the Royal Statistical Society. Series A (Statistics in Society)* 152 (1989): 11.
89 Philip Redfern, "Precise Identification through a Multi-Purpose Personal Number Protects Privacy," *International Journal of Law and Information Technology* 1 (1994): 313.
90 Redfern, "Population Registers: Pros and Cons," 11.
91 Redfern, "Precise Identification," 313.
92 Redfern, "A Population Register or Identity Cards," 506–507.
93 TNA, RG 50/23, Redfern to Hibbert, 30 Nov. 1990, 5.

5 The British People, Government Data and the 1984 Data Protection Act

The previous chapter pointed out how the government's data collectors were exasperated by the political opposition launched against the trialling of questions about race for the 1981 census. This attitude was not confined to those individuals involved in the census. In fact, it spoke to a more widely held set of reciprocally entwined views of the rational and beneficial nature of population data and the Luddite absurdity, or anarchic malevolence, of opposing this modernising data turn. A US sociologist wrote of the "wild" nature of such protests that, he argued, were based on a "sort of indifference or antagonism to science" and all it represented.[1] In a similar vein from Britain, Philip Redfern despaired of politicians who stood down in the face of these protests, most of which he regarded as a product of "irrational fears," regarding this timidity as little more than "political reflexes" rather than the product of any "rational discussion."[2] For their part, these politicians agreed that these issues aroused what were termed "strong emotional reactions" in the people and accordingly saw their own actions as characterised by caution rather than timidity.[3] Clearly tempers were running high on all sides of this issue. Yet gathering population data of necessity involved government entering into some sort of relationship with the people and this would, notwithstanding Redfern's obvious distaste for political reflexes, have to be structured through political action. But government had the upper hand here because, as Foucault noted (see Chapter 1), it was the government that took the initiative on population-data gathering and so, other things being equal, it could choose how to frame any congruent public political debate, or indeed whether to have any such debate at all. Its policy in this regard is the subject of the first section of this chapter.

The second section looks at the passage of the 1984 Data Protection Act and the subsequent formation of the Data Protection Authority (DPA) to indicate how, despite its name, this body was set up in such a way as to cause the minimum disruption to government plans to push on with widening, deepening and linking its population data. The third section discusses how, although it was set up to be a paper tiger, the DPA was able to use its new enclosure to raise some important issues. This section examines the registrar's work on credit reference agencies and National Health Service (NHS)

DOI: 10.4324/9781003252504-6

contract data sets. This details how, despite raising these issues, the registrar was too weak to institute the changes in legislation that were necessary to protect people from the creeping commercialisation of their personal information because, as it further shows, the government was at the forefront of these processes.

1

This book charts the course of a change in the attitudes of the British government towards population data to illustrate how, following 1964, the government came to see this information not just as something it needed to exercise its functions but rather, increasingly as what government was: it held data in order to pursue the biopolitical interventions that were the essence of modernity. Yet, from the government's perspective, while it was forging ahead, there was one thing that seemed to be a point of fixed reference in these processes: this was the British people. The population who were to be surveyed, assessed and normalised were ambivalent about these processes being, at best, all too easily swayed by rabble-rousers or, at worst, all their principled objections to data gathering and their stalwart defences of the traditions of Britishness could be little more than, as Philip Redfern phrased it, "a façade on the part of those who abuse freedom and cheat the community."[4] In either case, whether the people were principled or fraudsters, the result was the same: the government was never entirely sure it could trust the people to go along with a modernising agenda driven by data projects. Thus, the early pioneers of data-based public medicine and epidemiology found their plans ran up against public scepticism. Jerry Morris, for example, noted that: "a big change in the climate of public opinion will be needed for the growth of ... any public health approach to chronic diseases and it could happen only slowly."[5] Acheson outlined the attitudes that needed changing when he wrote that the unified system of numbering that he advocated

> has unpleasant associations with conscription, wartime security restrictions, identity cards and food rationing. Barely tolerable in the interests of national security in wartime, in peacetime a number is regarded with suspicion and hostility reserved for an instrument of the police state.[6]

Later, in 1989, when confronted a growing wave of hostility towards census-type data gathering, a United Nations seminar suggested that the root of these attitudes lay in the people's ignorance of how statistical operations worked rather than memories of totalitarian oppression.[7] But, whatever the cause, there was clearly something about the peoples' attitudes that confounded data collection.

In 1993, the annual report of the data registrar (see below) published research his office had undertaken to find out what the British people felt about personal information. This showed that when presented with a list of nine policy priorities, protecting personal privacy was third most important in 1991 (it attracted 73% of votes), being ranked on this list below preventing street crime (83%) and improving educational standards (78%), but above issues such as unemployment (70%) and protecting the rights of women (54%), for example. By 1993, personal privacy had been overtaken by dealing with unemployment as a priority (this was now prioritised by 78%), but still ranked fourth in the overall list with 66%. In 1993, 33% and 35% of people interviewed were either "very concerned" or "quite concerned" about the amount of personal information held by various organisations (the 1991 figures had been 40% and 32% respectively). When people were asked about the type of information they were most keen to keep private, personal financial information was placed highest by all surveys (information about savings and earnings were respectively chosen by 75% and 70% of interviewees in 1993).[8]

However, what really alarmed British people was not the idea that pieces of information, that ought to be private, could be exposed to view, rather they were concerned that the totality of their lives would be open to the gaze of someone in control of an interlinked system of data. This fear was encapsulated in a word that was commonly used in the earlier years covered by this book: dossiers. A dossier was not seen as a neutral file that opened the gateway to access government welfare, educational or health services, and neither was it imagined as the key to using the rights of citizenship. A dossier was seen as a malign compilation of subjective comments about the person concerned that was cached alongside details of the subject's life. Dossiers were feared as the inevitable outcome of any enhancement of government data gathering in general and of refinements to systems of data linking in particular.

Government may not have liked such attitudes, but it knew that these were not only the products of fevered imaginations played upon by rabble-rousing demagogues. Thus, in 1990, Philip Redfern admitted that though population registers (the ultimate linked data system) had a lot of "pros," they also bought with them a few "cons," one of which was that "the state is enabled to store and process vast quantities of data rapidly ... and to build dossiers unknown to the individual concerned."[9] The government also understood that these attitudes were current among the people. For example, in 1968, when the Wilson government was pondering its *People and Numbers* plan, the Homes Affairs Committee noted that: "people will fear that government, or even local, officials would have access to personal details of their lives," because the system would allow linkage of datasets to present such an overall picture. The committee then pointed out the nub of the problem: "the trouble is it is the possibility of linking information collected by different agencies that is seen as one of the most important advantages of the common number."[10] Since this conflict of interest was hardwired into any common numbering, it would remain an inherent feature of any discussion of the topic within government. Thus, the 1973 confidentiality guidelines produced by the Central Statistical Office (CSO) noted that people were "sensitive" about how their personal information was held and used and felt that they should have at least knowledge of, if not some control over these processes, adding that "a major cause of this concern stems from a belief that personal dossiers might be compiled and once compiled could be available for use against the interest of the individual."[11] Qualitative research conducted by the Office for Population Censuses and Surveys (OPCS) in advance of its plans to extend the use of the electoral register (see Chapter 6) fleshed out these conclusions. Thus, this research recorded that almost half the people surveyed showed some reservation about the scheme, whilst just short of a further fifth were hostile towards it. Here the interviewers recorded people as commenting: "in ten years from now they'll have complete case books on us all"; "privacy is being whittled away"; "I've always said that in ten years' time we will have a communist state here. I think we're well on the way now," and, "they can check up on you all the time. I don't want them knowing all about me and my family. It's my business, not theirs."[12] Additionally, about 6% of those surveyed were reported to be overwhelmingly negative about the proposals stating that they opposed virtually all government data gathering on principle.

All of this put the government's data gatherers in a quandary because though many people objected to this data gathering, a majority of them had, presumably, voted for the policies that it was used to implement. Moreover, from within the data enclosure, the view was always that the work of gathering and analysing population data was undertaken entirely "for the health and welfare of the nation." It was designed to provide wider benefits for the community in the form of "fewer calls upon the NHS and other services, less absence from work through sickness, improvement in the efficiency of the working population and so on."[13] From the perspective of the government, it was, in other words, a question of rebalancing public perceptions (see Chapter 2), and so its real dilemma focused on how it could execute this political balancing act.

One possible way to achieve this realignment was presented to the OPCS through its research into public attitudes regarding the planned extension of the electoral register (see Chapter 6). Here, researchers noted that if people were given more information about the ways in which their data might be used, they were less likely to object to its being gathered.[14] Moreover, such a realisation was not particularly new at this point (the research was conducted in 1977). In fact, the Department of Health and Social Security's (DHSS) enquiry into confidentiality procedures in 1971 had noted that: "attitudes and expectations about the use of personal records ... are not static."[15] Public attitudes were thus known to be malleable. Indeed, though this was precisely the point behind the accusations that Jeremiahs and Luddites whipped up public opinion, this pliability also opened the public up to being persuaded by voices from the other side of the debate. This was made plain by the fact that even some of the strong objections to the electoral register scheme, quoted above, were tempered by the use of phrases such as "I don't mind answering any questions as long as there is a good reason for it," or "I don't mind giving it [personal information] if I know what it's going to be for ... as long as the purpose is clear to me and I think it is a valid reason."[16] In other words, the public were open to persuasion, and attitudes could be changed through political dialogue. This was something Moser accepted. In fact, he insisted that:

> we are not powerless to try to make people feel more charitable towards statistics, more understanding about the need to complete returns, more rational about the internal use of information for compiling statistics. And we need to approach this in a more positive manner.

It was thus necessary, he argued, to work towards an "enhanced public awareness of the benefits accruing from good quality government statistics."[17] Qualitative research conducted for the office of the data registrar in 1989 came to the same conclusion. This noted that what emerged from its in-depth interviews with the public was a defensive state of mind, one "that appeared principally to arise from a sense of vulnerability, and its consequent feelings of powerlessness and lack of opportunity for active involvement in the investigation of information flow."[18]

One academic authority on government demography argued that the root of the problem experienced by the Harringay trials, in advance of the 1981 census, lay in the low priority given to public relations, since, at that time, the OPCS had only three public information officers, all of whom were employed only on temporary contracts.[19] However, the government did not address this staffing issue. Doing so would, for sure, have been problematic in the post-Rayner environment but, more importantly, the government never tried to engage the people in a serious political dialogue about the nature of its data gathering because it did not want to do so. This was because doing so would have exposed the nature of its plans and the whole thrust of the Government Statistical Service's (GSS) work, particularly after Rayner, though this was something that Moser (notwithstanding his advocacy of openness and dialogue) had long advocated, was towards the linking of departmental data sets. This ran contrary to the norms of British confidentiality codes and so was, the government realised, the very thing that the public most feared and was, therefore, likely to produce an outcry if exposed. Thus, the confidential report of the DHSS's 1972 enquiry into the confidentiality of records noted that, while the public might accept that the benefits of future research could outweigh some short-term infringements of data privacy, "it is not certain that this would extend to all research projects which involve the linking of sensitive records."[20] Consequently, open political dialogue was more or less ruled out as a means of achieving the rebalancing of the public's attitudes that the data turn required.

Nevertheless, the privacy campaign that had obliged Wilson to shelve the *People and Numbers* population register plan demanded some response from government. This was delivered in the form of the Younger Committee's enquiry into privacy established by the Labour government in January 1970 (it reported to the Conservatives in July 1972). This committee was very much a paper tiger and the government used it to deflect popular concerns about data gathering onto computer technology subsequently publishing white papers on this subject. Computerisation was gathering pace at this time and had been the object of some high-profile protests such as that made by the leader of the Liberal Party, Jeremy Thorpe, before the census in April 1971. On 14 April, Thorpe made a widely reported speech in which he promised to go to prison rather than answer some of the census questions because computers were to be used to process some of the returns. Thorpe objected to this because, he maintained, computers increased the accumulation and spread of information to such dangerous levels that they represented a threat to individual freedom.

The government did not want to stoke hostility towards its computers but the fact was that the technology was, through incidents such as this, already in the public eye, whereas the bulk of government data, which was still kept on paper, was not. The government understood that there was no essential difference between data held on computer tape, and data held in a filing cabinet. Thus, in writing to Margaret Thatcher, in 1989, Lord Young asked the prime minister "how can one justify making special rules for computer systems which do not apply to paper-based systems?"[21] While four years earlier Donald Acheson was told that the health department's legal advisors could, similarly, see no reason why any confidentiality code drawn up to cover personal health information should not cover both computerised and manual files. Indeed, the Medical Research Council's (MRC) guiding principles for all researchers argued that while it was "common practice to enter medical research data into information processing media … there is no essential difference between these and conventional records held on paper."[22] However, by accepting that computers had attracted hostile publicity and by further allowing, indeed by encouraging, attention to focus on this technology, government could effectively remove any real political discussion concerning the nature and purpose of the data itself and focus instead on the machines. This allowed government to offer a variety of technical solutions to what it encouraged people to perceive was the real issue at hand. In this manner, as Jürgen Habermas noted, "public discussions" were discouraged because they "could render problematic the framework within which the tasks of government action present themselves as technical ones."[23] Thus, for example, the government damped down Thorpe's protest about the 1971 census by inviting the Royal Statistical Society and the British Computer Society, to appraise the security of the census's computer systems. The two societies did so and the reports they produced gave the census security arrangements a clean bill of health. The government published these reports and Thorpe duly completed his census form.[24] By contrast, it is noticeable that when the protests in Harringay erupted over the 1981 census and targeted the motives behind the operation, as opposed to the way it was to be conducted, the government did not engage in any political dialogue with the community at all.

This increasing focus on how some data was stored, rather than on why any of it was gathered in the first place, shifted the terrain of debate to the extent that it can be seen as part of what Habermas has referred to as the "depoliticization of the population."[25] The series of processes involved in this depoliticisation have been defined by Matthew Flinders and Jim Buller as: "the range of tools, mechanisms and institutions through which politicians can attempt to move to an indirect governing strategy."[26] They are, however, careful to indicate that the term depoliticisation is something of a misnomer, since though the arena or processes through which decisions are made may alter, the politics remains. Indeed, they show that it is politicians who decide which policies will be depoliticised and what systems will replace the extant methods.

As the term depoliticisation is usually used, it suggests a rather narrow definition of politics, one that focuses on governing institutions and personnel. However, as Habermas uses it, the term is arguably more important since it refers to the removal of the people from the political process. As the term is used here, it denotes both these elements. Thus, government increasingly fell back on the language of statistical rationality to both justify what were fundamentally its political choices and to cut the people out of any debate on the subject of population data. It has been noted that the new terrain for depoliticised policy can be defined or justified in a variety of ways including the argument that "a single rational and technically correct solution ... exists."[27] This can work alongside views that rely on "scientisation," whereby the common form of rationality is so dominant that policy fields move beyond the stage of becoming depoliticised to become apolitical in the sense that there is no longer any debate about the means and ends of policy. The attitudes of the British governments reached this stage by the end of the period studied here. However, two things make it clear that this was the result of a political transformation.

First, this reframing of the questions around government data still left space for discussions concerning these means. In fact, as David Cope (he served on the OPCS's Advisory Committee on the 1981 census) noted, the government's solution to the public's concerns (as it redefined them) was to install technical safeguards on its computer systems and impose restrictions on the use of computerised data. However, the latter course of action tended to suggest that data had in fact previously been misused, while the former approach was beyond the understanding of many ordinary people, at a time when 40% of people had never even touched a computer and about a further 25% had only had the most cursory of interactions with the technology.[28] Hence in both cases, the door to further rounds of public suspicion was left open.[29] There would, in other words, always be people like Thorpe making protests like his. Second, though this depoliticisation was indeed the desired outcome for the government, this did not mean that it was always able to achieve a flow of data uninterrupted by serious political protest, Harringay again stands as the example here. Therefore, given the, somewhat fragile, nature of the government's achievements in keeping the lid on debate about its data operations in this period, politicians would, as Flinders and Buller note: "from time to time face pressures to ... justify their choice."[30]

Government pursued two strategies when it was placed on the defensive like this. The first was its more preferred method. This dealt with people as individuals and spoke to them as consumers of government services rather than engaging with them as members of a political community. Such an attitude was all of a piece with the wider ideological bent of governments that saw their mission as being to recast the people as competitive individuals: free operators in a market-driven environment removed from government interference. Thus, the government took the people's opposition to providing information at face value and accepted that they did not like being pestered to provide the same data to different bodies. However, in accepting this much, it went no further in analysing public political attitudes and thus gave a fillip to the processes of depoliticisation endemic to this data turn. It therefore did not even glance beneath the surface of this antipathy to understand that the people might tolerate this hassling if they were involved in the political discussions that established such processes in the first place.

After 1964, officials thus frequently remarked that one reason they wanted to remove the codes of confidentiality around government data was that these stood in the way of data sharing, and that such restrictive practices led to departments being "forced to duplicate questions in enquiries."[31] Similarly, Moser was not alone in claiming to want to save the people from "the burden of form filling."[32] Later, in the 1990s, the government's computerised innovations (see Chapter 8) were justified as representing something that would reorient government service provision towards the "customers" real needs and away from "form-filling, duplication and the need to deal with many different offices," this being a situation that government felt was "unsatisfactory and inconvenient."[33] In this manner, government's chosen, surface-level, reading of the public's attitudes gave it an excuse to do what it wanted: namely to link its data, while claiming that in doing so it was responding to the public's attitudes and giving them what they wanted.

The second method by which government sought to justify its data schemes was to mobilise the two issues examined in the previous chapter: fraud and immigration. Thus, when the Conservative home secretary Michael Howard seemed to be losing the political argument within government over his proposal to introduce ID cards, he wrote to the prime minister, John Major, arguing that the cards would be popular with the people as they would aid "both in preventing crime and in strengthening our immigration controls."[34] Following on from this, his 1995 green paper on ID cards trumpeted the scheme's ability to "guard against duplicate or false claims for state benefits" and to "make it more difficult for illegal immigrants to stay or to work here illegally."[35] However, these political points, referred to in the green paper as "wider benefits" to the individual, were dwarfed by the long list of "direct benefits to the individual holder" of the card.[36] In other words, though the government could provide some political justification for these schemes (albeit spurious and ill-informed, see Chapter 4), ministers and officials preferred to keep these to the fringes of any discussion since they always sought to centre these on the depoliticised, individual and consumerist-type benefits of their data-collection systems. It was in this manner, by accepting that only certain facets of the public's attitudes should be addressed, that these governments were able, to some extent at least, reposition and reformulate the terms on which they engaged with the public on issues around their drive for population data.

The overall purpose of the growth of the data state was not to empower the people as citizens of a democratic polity it was, by contrast, to study, normalise and protect them, which under these Conservative governments meant equipping them for their roles in a marketised society. This was much more likely to be achieved by means that were compatible with the ends: through addressing the public as individuals rather than as a polity. Government argued that the people wanted the best of both worlds: to be able to reap the benefits of population data yet still preserve the traditions of Britishness. Government sought to rebalance this relationship and a key element of this was to hold any discussion about its population data that it might be obliged to undertake, not with the community, but with individuals. This, of course, did not mean that the government always got precisely what it wanted when it wanted it. This was, after all, a political process and as such was open to the twists and turns imposed on it by political contingencies.

2

What the government wanted from political interactions with the people over its data operations, what it categorically did not want and what it would accept if it were obliged to, can all be seen in the course of events that led to the passage of the 1984 Data Protection Act. The privacy campaign that had stymied the revolutionary introduction of Wilson's population register led to the opening of an enquiry into privacy by a committee chaired by Kenneth Younger.[37] The committee's report recommended the formation of a data protection authority and a new committee to pave the way for this body was duly set up in February 1976, with Younger, once again, in the chair. However, he died three months later, and the committee was subsequently reformed under Sir Norman Lindop. Lindop reported in December 1978 and this report was left on the shelf by the Callaghan government (Callaghan had succeeded Wilson in April 1976), and remained there when Thatcher took office in May 1979.

The central section of the Lindop Report was Chapter 29, which discussed what the report termed "a universal personal identifier" (UPI)—in other words, the use of a common number to link files across all government departments.[38] The report suggested that there could be huge savings in administrative costs, along with an added convenience to the public, if all datasets referred to each member of the public by using the same number: a UPI. However, the greatest advantage to using a common numbering system like this would be that: "it would make it easier and cheaper, especially with the use of computers, to relate or merge the information about individuals held in different sets of records." Despite these advantages, the report pointedly placed itself on the side of the public, adding that the use of a UPI would, "also present a considerable threat to the privacy, and perhaps the freedom of, the private citizen" and would, moreover, "greatly reduce the British citizen's traditional anonymity."[39] It would, in other words, eviscerate the traditions of Britishness.

Notwithstanding this presentation of the public's darkest fears, Lindop was able to conclude that such an eventuality remained in the realm of speculation. This was because what Foucault described as Britain's "much looser, Anglicized system" was still intact.[40] Thus, Lindop's report stated that his committee's investigations did "not suggest that the UK Government has any plans at present to introduce either a population register or a UPI."[41] These had been "contemplated from time to time," Home Office witnesses had told the committee, but as the CSO chose to tell Lindop, the advantages of such systems "had always been outweighed by objections of principle."[42] The next chapter will provide detailed insights into what the government was actually engaged in at this time, which was very considerably different to what Lindop had been told. However, the main outcome of this campaign of camouflage was that it worked: it concealed the truth from Lindop, and the committee went ahead and made its recommendations apparently certain that its proposals were nothing but a codification of what was the current best practice across Whitehall.

On this basis, the report argued against allowing a UPI to be developed by stealth, for example, by a functional creeping of any extant set of reference numbers such as those used by the NHS. The report argued that the DPA should have the power to restrict the use to which any system of numerical identifiers could be put. The framework for the data linking essential to creating the public's nemesis, dossiers, should only be introduced, the report concluded, openly through legislation.

Had either Callaghan's or Thatcher's governments seriously wanted to allay the fears of the British people, then the Lindop report provided the recommendations by which to do so. That the report was left to gather dust thus indicates what the government's priorities really were. However, it also raises a question: given this desire to do nothing that would impede the gathering of population data, why did the Thatcher government pass the 1984 Data Protection Act?

The answer is that this government was obliged to do so because of changes in European rules. Britain was not the only European state to face protests over data gathering in this period and, in October 1980, in response to a growing wave of opposition to census gathering (which Redfern found "distinctly unnerving"[43]), the Council of Europe approved a *Convention for the Protection of Individuals with Regard to the Automatic Processing of Data.* This had two purposes: first, to encourage states to adopt data protection legislation and second, to ensure that such laws were compatible across signatories thus facilitating trade by reducing unilateral or restrictive measures. The convention was open for member states to sign from the start of 1981 and Britain signed in May of that year. However, the convention could not be ratified until domestic data protection legislation had been passed. Given this, any British government would have had to plot a tight course between two equally unpalatable alternatives. On the one hand, a policy that would damage the country's international trade and on the other, implementing something akin to Lindop to the detriment of its data-gathering missions. The result of this begrudging acquiescence to data-protection legislation was the 1984 Data Protection Act.

This act was described at the time as being little more than "a lot of holes stitched together."[44] These holes were usually seen as being the host of exemptions to the provisions of the act and the ease with which ministers could create these and shield them from public view. These exemptions were indeed important features of the act, but more important than these was the exemption that was spun into the warp and weave of the legislation. This was that it only applied to data held on computers; it did not apply to paper-based systems. With the benefit of hindsight, this focus on computerisation may seem well targeted. Indeed, on this basis, a recent overview of surveillance in Britain presents the 1984 act as defying the trajectory of Orwell's *1984* precisely because it applied to computers.[45] But no one involved in drafting the 1984 act seriously wanted to limit governments in the future. Far from it, what they wanted to do was to free government in the mid-1980s from any oversight of its population-data operations, and since these were predominantly conducted on and through paper-based systems, a focus on computers would have had this effect. In this way, the Data Protection Act did not protect the British people in 1984 from government data, it protected the government's data, as it stood in 1984, from the people.

The government was, of course, well aware of the equivalence between its paper and computerised systems with one Home Office official noting simply that any differentiation drawn between the two did not rest on considerations of "logic."[46] Moreover, when ordinary British people found out about this distinction, they could not see how it made sense either. Research conducted for the data registrar established that people found the idea that they could only use the DPA to find out about information that was stored electronically to be "strongly disappointing."[47] While the year before, in his report for 1987, the registrar noted that: "a large majority also favoured legislation to control the information that may be kept … on paper as well as on computer."[48] Moreover, this attitude was also widespread within his office. Thus, in early 1988, when members of staff were asked to complete an internal comments sheet to list problems they had encountered in administering the 1984 act, and any solutions they might be able to think of to address these, a large number of them wrote, amongst other things, that: "the act should be extended to cover manual files."[49] The deputy registrar concurred with these views and, in 1987, in an internal report monitoring the administration of the act, placed first on his list of problems the topic of "scope and objects of the legislation" where he wrote, in staccato note form "not manual records; all data, however non-sensitive … if it happens to be electronic, is this … a confidence trick, or what?"[50]

3

The office of the data registrar had only been set up at all because the government had been backed into a corner by European legislation, and once in this position, it recognised that it had to walk a tight line. If it imposed too onerous a set of restrictions on the DPA, the office would lack all credibility, so it set the office up with as tightly circumscribed a working brief as it could and hoped that it would be able to "rely on the registrar's good sense."[51] In this, the government may have hoped that the registrar would conform to the norms of what Michael Moran has described as the traditional British "club" system of regulatory bodies, where regulators and the regulated were all drawn from the same political-cultural milieu.[52] Nevertheless, just as Moser had established the OPCS to form the data atmosphere for government, but then found it moved somewhat beyond his control, so the government found the data registrar, Eric Howe, used his enclosure to challenge its data turn. These challenges came because, as has been seen, the new office held the attitudes towards data linkage that were common amongst the British people. This section analyses two of the arguments that developed between government and the registrar. The first of these concerned credit reference agencies.

Credit reference agencies were in many ways the public face of the shifts in data gathering and use that became increasingly prevalent in this period. As such, they were also the British public's bête noire. Had the public been aware of how extensively the government was harvesting their information and forming it into an extensive matrix, then, as the government was well aware, the focus of this ire might have shifted. However, as things stood, the registrar's research found that when people were asked which institutions they trusted to use their personal data in a responsible way, credit reference agencies were at the bottom of the list of ten choices, being trusted by only 24% of those surveyed in March 1988. The people who responded to this survey placed these agencies below doctors and the NHS (trusted by 89% of those surveyed), banks (80%), the police (69%), employers (63%), the tax authorities (57%), schools/colleges (56%), the DHSS (52%), shops (35%) and mail order companies (25%).[53]

Credit reference agencies kept files on members of the public and, by 1988, the various agencies were discussing unifying their systems to produce a national credit register. Their systems included publically available information, such as county court judgements, along with details supplied by the lenders who used their services. However, they also based their assessment of an individual's creditworthiness on information pertaining to other people resident at the same address. These could be other family members, lodgers or even former occupants of the property. These details could, depending on the policy of the agency concerned, also include all occupants of an applicant's previous two addresses. But precisely how many other people's records were searched in order to assess an application for credit was, as far as the registrar was concerned, irrelevant: using other people's information without their consent violated their privacy and as such was in contravention of the terms of the 1984 act. Moreover, the length of time this information was held on file, virtually permanently, constituted another breach of the act. The registrar's office wrote to the credit agencies telling them of his dissatisfaction with their practices, suggesting that he would give them time to introduce new methods, proposing that they meet to draw up a plan of action and reminding them that, though he was holding them in reserve for the time being, he held enforcement powers and would be prepared to use them.[54] Were the pressure for a national credit register to have come to fruition, it would have produced a system that would have held records "on the vast majority of adults in the United Kingdom."[55] Thus, the registrar was concerned to make sure that, were it developed, it would be legal and conform to public expectations of fairness.

The government had "heard rumblings" about this gathering storm for some time prior to this suggested meeting because Treasury officials were kept fully informed of the registrar's moves by the credit industry's representatives in the Finance Houses Association (FHA).[56] The FHA was particularly concerned by the registrar's proposal that people should be the legal owner of data pertaining to themselves and that this should not be shared without their express permission.[57] Given that records showed that a third of applicants were less than truthful on credit applications, the best that could be said for the registrar's proposals, according to the FHA, was that: "they display little understanding of the credit industry," while in the worst-case scenario they could, "effectively destroy point-of-sale credit as well as exacerbating bad debts and arrears." Thus, giving people this ability to opt out would, the FHA warned, "destroy the whole system."[58] Arguing against this apocalyptic vision, the registrar showed that over two thirds of the people who responded to his office's surveys had stated that they were either "quite concerned" or "very concerned" about the amount of information held on them by others.[59] Moreover, it was (as mentioned above) information about financial matters that the public most wanted to keep private.

Eric Howe realised that a national credit register was very unlikely to appear without a government endorsement.[60] However, he may have been unaware of the full extent of the support the credit industry could muster across Whitehall. Here, a meeting including representatives from the Bank of England, the Department of Trade and Industry and the Treasury, was held in order to discuss ways to "convince" the registrar "of the wider public policy issues involved in credit referencing, and the importance of maintaining a credit reference system which minimises arrears and bad debts."[61] It was the policy of all these bodies to encourage the use of credit referencing and ministers from the Department of Trade and Industry were "on record as saying that on this issue questions of privacy were outweighed by lending considerations."[62] The meeting discussed whether Treasury ministers could contact their colleagues in the Home Office, which had departmental responsibility for the registrar's appointment and budget, in what was clearly an effort to put crude pressure on Howe. This came to nothing however, when these officials realised that the registrar was legally independent and reported to Parliament. After this avenue to changing the registrar's attitude was revealed to be a dead end, the officials' main hope lay in getting Howe to accept what the director general of fair trading presented as a compromise position. Under this proposal, the reference agencies would limit their use of information on third parties to include only family members and would exclude anyone else at the same address or previous occupants of it, when assessing applications. However, this solution entailed Howe making a quid pro quo move of withdrawing any proposal to give people a veto over the way any data about them could be used.[63] So what this really involved was the agencies abandoning part of a set of practices that were to all intents and purposes in contravention of the terms of the 1984 act, while the registrar would give up pursuing the people's rights to have the final say over how their data was used, which was a principled cornerstone of the act.

Understandably, Eric Howe rejected this so-called compromise and his report of 1989 noted that the number of complaints his office had received about credit reference agencies had doubled over the previous year, which presumably stiffened his resolve. The previous autumn the chancellor, Nigel Lawson, had received an update on progress towards the establishment of a national credit register. The government supported this and Lawson was told of moves by a variety of banks and building societies towards implementing this national scheme. Banks were restricted from sharing information on their customers by law, but this law, Lawson was reminded, was currently under review. Moreover, while the banks waited for the findings of this review, a system was being trialled for sharing information between building societies and credit reference agencies. The data protection registrar remained opposed to these developments and so, Lawson was told, credit agencies might look to him for support in this conflict, with his officials adding: "we welcome these moves."[64]

By the summer of 1990, Eric Howe's office had clearly had enough of the credit industry's prevarication, and enforcement notices were served on the main agencies. The case of that served on Infolink will serve as an example of these here. This opened by stating that discussions between the registrar and the company had dragged on since May 1986, it listed in exasperated detail the steps taken by the registrar to change the company's practices before concluding that: "no substantial progress has been made."[65] Infolink lodged an appeal against the enforcement and this was heard in April 1991. This decision implemented a system that was more or less the same as the position brought forward by the director general of fair trading, which had been previously rejected by Howe. Thus, in assessing an applicant, the credit reference agencies would no longer be allowed to use information on previous occupants of the applicant's address, or previous occupants of any prior address the applicant gave, neither would they be allowed to use information pertaining to a current resident of the applicant's current address with a different family name (a lodger, for example), but they would be able to use information about people currently residing there who had the same, or a similar, family name.

The most important point about this decision, and the plan designed by Whitehall that it clearly followed, is the extent to which it overrode even the scant protection offered to the people by the 1984 act. Though the agencies gave up their practice of surveying swathes of people when assessing an applicant for credit, they were still permitted to assess members of the applicant's family or people who could be mistaken for family members without asking these people for their permission to do so, all of which was to permit the agencies to circumvent the terms of the 1984 act. Though the government wanted a national credit register, it did not want to have to legislate to get one and thus expose itself to charges of building financial dossiers on the people. In this sense then, control over the data on which credit relied was best left to the credit reference agencies, since they would be the users of this resource. The 1984 act and Eric Howe stood in the way of this drive and so were overruled.

Towards the end of November 1992, John Major succeeded Margaret Thatcher as Conservative prime minister. By then, the Thatcherite agenda of privatisation and marketisation was well established and this policy vector was maintained by her successor. This programme had already had an impact on the NHS and, in February 1993, the data registrar turned his critical eye to the consequences of this for the use of data in the health system. The result of this investigation formed the second challenge the registrar posed to the government's data turn to be considered here. This investigation was founded on a complaint received from the Joint Consultants' Committee. These senior medical practitioners were concerned about the amount of data included in the system's minimum data sets and particularly the fact that these held patients' names.

Since 1979, the government's reforms had divided the NHS into providers and purchasers of health care. The former group included hospitals and corporate bodies called NHS trusts that were controlled independently of district health authorities (DHA). In cases where all the provider bodies within a DHA had become trusts, the DHA would remain as a purchaser of health care. In other cases, it might contain both provider and purchaser bodies. However these bodies were managed, contractual-type arrangements existed between them and, when services were provided, patient data moved between them. This was the minimum data set. This information moved between groups of health professionals and managers who were operating in a system that placed them at arm's length from direct government control: this being the whole purpose of the Thatcherite reforms. Nevertheless, the Department of Health, through its NHS Management Executive, remained responsible for "setting standards, providing guidance and placing requirements on individual data users."[66] The minimum data sets that were compiled and transferred through the system included data under four main headings: details of the contractual parties, details of the patient, data on the administrative side of the patient's care and information about the clinical facets of the care provided. The patient's name, address, ethnicity and marital status were all recorded, under the heading "patient details," and it was these pieces of information that drew the registrar's attention. The inclusion of the patient's name meant, of course, that the person concerned could be identified by anyone handling the data set which thus created the possibility of a breach of that person's privacy. An additional point of concern arose because neither the registrar nor the consultants, who made the original complaint, could understand why it was necessary for anyone to know these other pieces of personal information. Moreover, neither the providers nor the purchasers seemed to actually use this data for any administrative purpose.

The Department of Health claimed that it was useful to know whether patients were married as this could indicate the level of support that they might receive upon discharge from a hospital and that holding information about patients' ethnicity could aid research. The registrar pointed out that being married did not necessarily mean that support was available. Moreover, though he accepted that the collection of data on ethnicity for research purposes could be acceptable, this was only legitimate if such research was clearly being planned, adding somewhat caustically, three times, that: "it is not acceptable … to hold data on the basis that it might come in useful one day."[67] These were important bones of contention, but the biggest issue the registrar raised in connection to all this was that the patients at the centre of these minimum data sets were clearly identifiable.

As was noted in Chapter 1, Donald Acheson sought to turn the NHS into a data system where all information was centralised, standardised and linked and so would stand ready to be used by epidemiological researchers. Howe agreed that such research was important and that some might only be possible if there were accurate patient identification. Additionally, he accepted that it was important to connect the various events in a patient's healthcare history in order to understand that person's needs. He agreed that names could not serve to provide such a linking device as they could change over time or be given or recorded in varying forms. Therefore, to improve accuracy, it was necessary to add to the data, by giving and recording something such as a patient's date of birth, for example. However, here he made two important points. The first was that though such identification/linkage might be necessary for some projects, it was not necessary for all research and so this information ought not to be recorded as a matter of course. Moreover, his investigations revealed that though such recording was the norm across the NHS, not all providers actually linked this information.[68]

Howe's second point concerning this data linkage was broader and more important. Thus, he emphasised that the more data that was recorded for identification purposes, the more vulnerable the patient became. Chapter 2 showed how, in the early 1970s, the traditions of British data confidentiality were stretched very considerably by the use of the term "task" to define who could access personal information. It also showed how when the use of this term was instigated, its advocates saw its advantages as coming from its "fuzzy" nature and here, in 1993, Howe came up against the reality of what this vagueness meant for NHS patients. The 1984 act stipulated that data could only be held if it had been fairly obtained and Howe did not think that this was the case with what had clearly become NHS practice. This was so because, whilst patients would almost certainly accept that their personal information should be shared with everyone involved in their care, the registrar's research had shown that it was highly unlikely that they were aware of how far this information was actually being spread, or that it was being used for purposes unconnected to their immediate care. Thus, the report concluded that: "such uses and disclosures for other purposes including those associated with the data sets, should be explained to the patient at the time the information is collected."[69]

Thus, what Howe wanted was for people to be seriously engaged in meaningful political dialogue about the reasons for data gathering. This would have conformed to what people stated they wanted in the surveys discussed above (for example: "I don't mind giving it if I know what it's going to be for ... as long as the purpose is clear to me and I think it is a valid reason."). However, as has been seen, such engagement was not the government's preferred option. Rather the government favoured technical solutions to this issue and chose one that would not only stymie dialogue, but would also boost its own data operations. In doing so, it followed the trail laid by Acheson.

Acheson's solution to many of Howe's worries would have been to further the use of the NHS number, which would conceal the patient's identity. Howe agreed that the use of the number here would have had the desired effect; however, this raised an even more serious issue. By 1993, the year of this report on NHS contract data sets, the NHS number existed in a variety of formats and a decision had been taken to standardise this and issue people with a new number in a revised, uniform format. Howe was very concerned that this number, which would have near universal coverage of the population, should remain "context specific": in other words, that its use should be strictly limited to the NHS. His concern was thus that this new number could become a de-facto national identification number, the UPI that Lindop (above) had seen as a threat to privacy. The NHS Management Executive had made it plain to the registrar that they wanted the use of the new NHS numbers to be limited to the health system but, while he had "no reason to doubt the Management Executive's intentions," Howe found that he had to conclude that the executive "will not necessarily be in a position to restrict the use of the new number to the NHS." Thus, where Lindop had been hoodwinked by the platitudes of Whitehall, the registrar saw clearly that the NHS authorities would be subject to "pressures which seem likely to arise for the wider use of what will be a well-maintained, actively-used, unique national identification number."[70] Nevertheless, the registrar reached the same conclusion as Lindop. Indeed given the way his role had been configured, this was the only conclusion that he could reach, namely that statutory provision should be introduced to ring fence use of the NHS number. In other words, he was left to conclude that the government should police itself.

The registrar's concerns with both of these issues, the NHS contract data sets and the activities of credit reference agencies, were grounded in wider concerns about the commercialisation of data. When the House of Commons Home Affairs Select Committee first examined the work of the registrar in December 1990, just after Howe's office had served enforcement notices on the credit reference agencies, the topic of trading in personal information was high on their agenda. Howe outlined to the committee how data trading worked: pointing out that a customer who supplied information to facilitate a purchase "may then be bombarded with information on a whole range of other goods and services completely unrelated to the original purchase."[71] The committee agreed with Howe's judgement that people ought to be told, when they provided this information, who would see it in addition to those it had been given to and why these other parties would have this access. The system in use at the time was that a person in this situation had to go to the trouble of opting out of this sharing after the system had been initiated. The committee thought that this put things in the wrong order and prioritised the power of the data gatherer rather than that of the person who owned this data. Therefore, the committee's report recommended that: "the government consider amending the act to specify that an organisation must give broad details to anyone submitting personal information as to the future uses of that information."[72] There is no record of the government considering this issue, but had it done so, in transferring its contract data sets, the NHS would have been in clear breach of the law since the issue of informing people how their data would be used and distributed was essentially the same for the NHS as for the hypothetical shopper left to fend off a barrage of junk mail discussed by the committee.

In fact far from attending to this issue, the government was content to watch it grow. Thus, the report from the data registrar for the following year (up to June 1992) commented on the increasing pressure placed on public information files. This ratcheting up of the force applied to public bodies came at least partly from the pressures they faced from the government to embrace commercialisation. The report gave the examples of how the BBC had misused its lists of people that had bought or rented a TV set and how the Driver and Vehicle Licensing Agency had sent out information about a driving school along with provisional driving licences. The following year saw *The Sunday Times* run a series of exposés of the growing market in personal data. The newspaper alleged that this information was increasingly available to those who sought to buy it, with the sources listed including government departments. The registrar's report for the year to June 1993 concluded that it was highly likely that these press reports were credible and cited as evidence the behaviour of credit reference agencies. This report went on, like the report of the Home Affairs Committee two and a half years earlier, to recommend that the government should tighten the law. By this point, in 1993, the government was prepared to accept that some alteration in the law was necessary perhaps because it could not actually deliver any such change at that point. This was because a European data directive was being discussed and this process (see Chapter 9) had drawn down a moratorium on any reform of UK law.[73]

While the registrar had inklings of what was going on from complaints he received and from press reports, the government had a much clearer picture. For example, in March 1990, Diana Walford, the deputy chief medical officer, wrote to the MRC giving her views on the latest iteration of the government's confidentiality guidelines and, as part of this, she mentioned an incident that had recently come to light. This concerned private companies that had offered to give computers to GPs in return for access to some of the data that would be held on these systems. This could all be presented as a lapse of security, something that was thus open to a technical solution, but in reality, this was an inevitable outcome of the drift of government policy. The Thatcher government had liberalised the credit industry increasing the personal debt held by British people from nine billion pounds, in 1981, to twenty-seven billion pounds in 1987; in such circumstances, it needed credit reference agencies. Similarly, it had introduced an internal market into the NHS, and markets trade on, and in, information. In 1988, Pickford's report on the GSS referred to population data as the "fourth resource" of a modern economy and the Thatcher/Major governments certainly sought to increase the exploitation of this resource.[74] Some of this was not done directly by government, but rather by market-driven enterprises put into motion by government policy. However, other aspects of data use, campaigns against scroungers and immigrants as part of a biopolitical project to turn the British into a people capable of standing unsupported by the state against the rigours of a free market, clearly emanated from government itself. However, regardless of how these policies were executed, they left an increasingly depoliticised community, devoid of the traditional protections of Britishness in their wake. More of the policies government pursued in achieving these ends will be examined in the following chapters.

Notes

1 Harvey M. Choldin, "Government Statistics: The Conflict between Research and Privacy," *Demography* 25 (1988): 153.
2 Philip Redfern, "A Population Register or Identity Cards for 1992?," *Public Administration* 68 (1990): 506; and, TNA, RG 22/54, Redfern to Pennington, 3 Dec. 1986, 2.
3 TNA, HO 328/130, Homes Affairs Committee, *Population Registration and Common Numbering System*, 23 Oct. 1968, 3.
4 Philip Redfern, "Sources of Population Statistics: An International Perspective," *Population Projections: Trends, Methods and Uses* (London: OPCS Occasional Paper, 38, 1990), 107.
5 TNA, FD 12/1637, Medical Research Council, *Review of the Field of Social Medicine*, 9 Apr. 1963, 2.
6 E. D. Acheson, *Medical Record Linkage* (London: OUP and the Nuffield Provincial Hospitals Trust, 1967), 104.
7 TNA, RG 19/894, U.N. Statistical Commission and Economic Commission for Europe, *Seminar on the Relevance and Importance of Population and Housing Census Data*, 24 May 1989, 8.
8 The Data Protection Registrar, *Ninth Report of the Data Protection Registrar, June 1993* (ondon: HMSO, 1987), 95–97.
9 Redfern, "Sources of Population Statistics,"106.
10 TNA, HO 328/130, Home Affairs Committee, *Population Registration and Common Numbering System*, 23 Oct. 1968, 3.
11 TNA, RG 19/758, *Confidentiality Guidelines*, Feb. 1973, 9.
12 TNA, HO 411/59, *Report on an Exploratory Enquiry into Public Attitudes Towards the Extension of the Electoral Register System with Proposals for the Main Survey*, 17 Jan. 1977, Appendix A, 3–4.
13 TNA, RG 20/167, *An Examination of the Department's Policy of Charging Social Research Workers for Special Work Undertaken on Their Behalf*, 16 Jan. 1964, 2.
14 OPCS, *Extending the Electoral Register – 2, Two Surveys of Public Acceptability* (London: OPCS Occasional Paper, 21, 1981), 17.
15 TNA, BN 102/5/2, Working Party on the Confidentiality of Personal Records, *Revision of First Draft of General Principles*, n.d., ca., Nov. 1971, 2.
16 TNA, HO 411/59, *Exploratory Enquiry into Public Attitudes*, Appendix A, 3–4.
17 TNA, PREM 16/1884, Moser, *Review of the Government Statistical Service*, 4 Aug. 1978, 36 and 53.
18 TNA, JX 11/11, *Qualitative Research Conducted to aid Development of Advertising for the Data Protection Act*, Aug. 1989, 7.
19 Choldin, "Government Statistics," 148.
20 TNA, BN 102/4, D.H.S.S., *Confidentiality of Personal Records, Report of the Working Party*, Apr. 1972, 24.
21 TNA, PREM 19/2717, Young to Thatcher 6 Apr. 1989, 2.
22 TNA, FD 7/2336, Gowans to Acheson, 6 Mar. 1985, 2; and TNA, FD 7/2336, M.R,C, *Responsibility in the Use of Personal Medical Information in Research, Principles and Guide to Practice*, 1983, 9.
23 Jurgen Habermas, *Towards a Rational Society* (London: Polity, 1987), 103–104.

24 Peter Gerald Moore, "Security of the Census Population," *Journal of the Royal Statistical Society* 136 (1971) 583–596.
25 Habermas, *Towards a Rational Society*, 103–104.
26 Matthew Flinders and Jim Buller, "Depoliticisation: Principles, Tactics and Tools," *British Politics* 1 (2006): 295–296.
27 Ibid., 308.
28 The Data Protection Registrar, *Third Report of the Data Protection Registrar* (London: HMSO, 1987), 40.
29 David R. Cope, "Census-taking and the Debate on Privacy: A Sociological View," in *Censuses, Surveys and Privacy*, ed. Martin Bulmer (London: MacMillan, 1979), 186.
30 Flinders and Buller, "Depoliticisation", 296.
31 TNA, CAB 139/751, Kershaw to Fry, 6 Mar. 1970, 2.
32 TNA, RG 19/758, *Confidentiality Guidelines*, Feb. 1973, 8.
33 TNA, CAB 202/43, *The Government IT Strategy, Annex E*, 28 June 1996, 2.
34 TNA, PREM 19/4735, Howard to Major, 21 Sept. 1994, 1–2.
35 *Identity Cards: A Consultation Document*, Cm. 2879 (May 1995), 31.
36 Ibid., 11.
37 See Kevin Manton, *Population Registers and Privacy in Britain, 1936–1984* (London: Palgrave MacMillan, 2019).
38 *Report of the Committee on Data Protection*, Cmnd. 7341, Dec. 1978, 260–264.
39 Ibid., 261 and 264.
40 Michel Foucault, *The Birth of the Clinic* (London: Routledge, 1989), 48.
41 *Report of the Committee on Data Protection*, Cmnd. 7341, Dec. 1978, 263.
42 Ibid., 263.
43 Redfern, "A Population Register or Identity Cards?," 506.
44 Duncan Campbell and Steve Connor, "The Battle against Privacy," *New Statesman*, 9 May, 1986, 15.
45 Rhodri Jeffreys-Jones, *We Know All About You: The Story of Surveillance in Britain and America* (Oxford: Oxford University Press, 2017), 167.
46 TNA, HO 287/2353, Cairncross to Fitall, 10 Apr. 1980, 2.
47 TNA, JX 11/11, *Qualitative Research to aid Advertising for the DPA*, 7.
48 The Data Protection Registrar, *Third Report of the Data Protection Registrar* (London: HMSO, 1987), 18.
49 TNA, JX 11/2, *Monitoring and Assessment of the Act*, see for example, return by Bill Palmer, 9.
50 TNA, JX 9/2, *Monitoring and Assessment of the Act, Report of the Deputy Registrar*, 11 Dec. 1987, 2.
51 TNA, CAB 164/1694, *Data Protection: Transborder Data Flows, Meeting*, 16 July 1982, 2.
52 See, Michael Moran, *The British Regulatory State: High Modernism and Hyper-Innovation* (Oxford: OUP, 2003), 179–183.
53 The Data Protection Registrar, *Fourth Report of the Data Protection Registrar* (London: HMSO, 1988), Appendix 9, Figure 9.2.
54 TNA, T 555/306, A*genda, Meeting on the Supply and Use of Credit Reference Information*, 19 May 1988.
55 TNA, T 555/306, E. J. Howe, *A National Credit Register? Channeling the Tide*, 15 Feb. 1988, 2.
56 TNA, T 555/306, Wiseman to Hall, 10 May 1988, 1.
57 TNA, T 555/306, *Howe, A National Credit Register?*, 4.
58 TNA, T 555/306, Wiseman to Hall, 10 May 1988, 3.
59 TNA, T 555/306, Howe, *A National Credit Register?*, 2.
60 Ibid., 1.

61 TNA, T 555/306, Wiseman to Hall, 10 May 1988, 4.
62 TNA, T 555/306, Hall to Mackenny, 12 May 1988, 2; and, TNA, T 555/306, *Credit Reference Agencies and the Data Protection Act, Meeting*, 13 May 1988, 2.
63 TNA, T 555/306, *Address by the Director General of Fair Trading: Credit Reference Agencies, Third Party Information*, 19 May 1988, 4–5.
64 TNA, T 555/306, Sharples to Chancellor, 5 Oct. 1988, 4.
65 TNA, JX 10/3, *Enforcement Notice: To Infolink ltd*, 28 Aug. 1990, 7.
66 TNA, JX 2/10, The Office of the Data Registrar, *NHS Contract Minimum Data Sets*, Feb. 1993, 24.
67 Ibid., 11, 9 and 10.
68 Ibid., 12.
69 Ibid., 17.
70 Ibid., 19.
71 Home Affairs Committee, Session 1990–91, *First Report, Annual Report of the Data Protection Registrar*, 12 Dec. 1990, viii.
72 Ibid.
73 The Data Protection Registrar, *Ninth Report*, 10–11.
74 TNA, CAB 164/1910/1, *Scrutiny of Government Economic Statistics*, 27 Sept. 1988, 10.

6 Data Systems 1979–97 (1), the Electoral Register, the Poll-Tax Register and Data Linkage

As was seen in Chapter 5, the Rayner review accelerated trends, such as the centralisation of the British state's data systems, that predated the review itself. By the early 1980s, the Office for Population Censuses and Surveys (OPCS) recognised both the need for data and the shortcomings of existing systems but was also fully aware of how Rayner had insisted that it drastically cut its operations. To square this circle, it turned to an increased use of what systems and datasets were already at its disposal. The first section of this chapter highlights how this led to plans to repurpose the electoral register. The second section of this chapter looks at the registers that were at the centre of the poll tax (the community charge) a flagship policy introduced by the Thatcher government in 1987. This tax was built around registers of those liable to pay and, since these potentially included all adults, they were a possible source of data to be used in lieu of a population register. Both the electoral and the poll-tax registers reflected a long-standing characteristic of Britishness in that both were held locally, rather than centrally, and these two sections of this chapter examine how the centralisation inherent in the data turn impacted on this localism. The third section investigates how these governments continued to try to realise Acheson's blueprint for making more thorough use of the data already held in disparate filing systems across Whitehall. This demonstrates how government pursued a variety of schemes to link datasets through common numbering systems, a policy that became increasingly the data gatherer's metier despite the fact that, as the previous chapter showed, this was exactly what the British people did not want to happen with their personal information.

1

As was mentioned, in Chapter 3, in order to utilise more effectively the population data resources that were already at the government's disposal, the OPCS drew up plans to substantially extend the form and use of the electoral register. It also (see Chapter 5) conducted studies on how such a system would be received by the public. Though this proposal was drawn up, researched and rejected in the mid-1970s, it was published, in 1980, against the backdrop

DOI: 10.4324/9781003252504-7

of Rayner's work. At this point, it was seen by its proponents as a way to circumvent the impacts of the government's view "in which the effective limitation of public expenditure has assumed paramount importance as a national objective."[1] This section explores this proposal in more detail.

The project was called *Extending the Electoral Register*, but its subtitle, *extending the electoral registration canvass to provide better population statistics*, made it clear that this project was part and parcel of the thinking on population data that had burgeoned within government since 1964. Thus, this project began from the same conceptual starting point as all the others studied here, regardless of when they were launched: the view that all existing data streams were insufficient. In fact, the report maintained that it was "universally accepted" that the census, always the lodestone of population-data systems, became less useful with the passage of time from the year that it was taken. For example, "by 1971 the error in the estimate for Liverpool reached a disastrous 9% and for some smaller authorities it was even greater."[2] All of which was compounded by the decision to abandon the 1976 mid-term census.

None of these generic criticisms meant that the OPCS could simply use the existing electoral registers in place of the census, since the fact that they did not count the whole population meant that these were not up to that role. In fact, the registers did not even enumerate the whole electorate. The report showed how a 1971 comparison of the electoral registers and census returns revealed that the voter rolls were deficient by between 5 and 10% in forty-two English constituencies, and by between 1 and 5% in a further one hundred and seventy-four areas, with the two sources only coinciding in eighty-seven constituencies.[3] However, there were more serious structural impediments built into the nature of the electoral registers that meant they could not replace or augment the census without being considerably reformed. Thus, the electoral registers only listed people over eighteen; they did not classify people by gender, age or marital status; they did not hold information on resident foreign nationals, regardless of their age; they did not provide information on previous addresses of the people listed and they did not give details of separate households living at the same address.[4]

Incorporating all this into the annual canvas undertaken to update the electoral register would, the report stated, be "a major change in principle in the role of the electoral register."[5] But even this was an understatement because what this proposal would have produced would have been a population register in all but name. The report made this plain when it admitted that for the scheme to work as envisaged (to produce "statistics of uniformly high quality which are conceptually comparable between areas"[6]), it would need legislation to both require local authorities to execute the canvas, and to require people to complete it. It would, in other words, need to be compulsory.

The OPCS was aware that this plan might produce a popular backlash, and with this in mind, it forwarded two suggestions. The first concerned the timing of the introduction of a redesigned electoral canvas. The OPCS did not want this plan introduced immediately because this might endanger the 1981 census and it had already (at Harringay, see Chapter 4) had its fill of threats to this gold standard of its systems. Second, there was the more vexed question of what this data might be used for. On this point, the OPCS compromised and suggested that the data gathered would not become part of any central bank of information to be used for administrative purposes. Thus, though it would be, in all but name, a population register, it would not be used as a central clearing house for all government departments, rather its contents would remain confidential within the OPCS.[7]

In Chapter 3, it was shown how this scheme to repurpose the electoral register was killed off, in the mid-1970s, by Stella Cunliffe's appraisal of how much it might cost. However, in the 1981 report, the OPCS was keen to show how this scheme could actually save money. Population data was used in assessing the Rate Support Grant paid by central government to local authorities. Local authorities regarded a miscalculation of as little as 1% in the population estimates that underpinned this grant as unacceptable, while the OPCS recognised that errors of the order of 2% could have serious financial implications for councils. However, one of the main reasons the OPCS wanted this improved data stream was to assess the movement of people round the country and just as some local authorities would find their populations increasing, as a result of immigration, so others must have found theirs decreasing. Using a reliable measure of this movement would thus enable central government to make accurate, targeted payments and so proffered the chance of saving the money that might otherwise have gone out in overpayments.[8] It might, with such reasoning, be possible to make this population data system appeal to central government. However, this population data was also of great importance to local authorities, particularly in times of austerity driven by a central government that was politically hostile to the leaderships of many of the larger, more prominent, metropolitan authorities across Britain.[9]

By the 1980s, many local authorities had started to build their own data gathering systems. These were often developed to overcome either perceived deficiencies in central government information, or the complete absence of any such data. This latter was seen to be the case by West Yorkshire Metropolitan County Council that suggested to central government that the 1947 Statistics of Trade Act needed to be amended to permit the release of more economic data to local government.[10] This authority also saw that there were "serious shortcomings" in the census data provided by the centre. The authority's neighbours in South Yorkshire held similar opinions and presented a catalogue of errors and omissions in information from central government, along with lists of types of data that were formatted in different ways making it difficult, at best, to draw comparisons across data sets and geographical areas, all of which was echoed at the 1980 Strategic Conference of County Councils.[11] Moreover, writing from the epicentre of government data, on the cusp of Thatcher's accession to power, Moser agreed with these descriptions of the shortcomings of central government's data offerings to local authorities.[12]

Given this situation, local authorities did what they could to plug the gaps in their own knowledge systems. The West Yorkshire Metropolitan Authority developed its own data system, which combined statistical data and information garnered from the local press and any other sources at its disposal. In 1974, Hertfordshire County Council began operating a system that used punched cards. Here a summary of information gathered was held on a card that was perforated to indicate key points of classification such as policy area, location or date. By 1979, this council's system held over 6,000 cards and the system was groaning under the weight of information to such an extent that the council was looking to computerise it all.[13] On this point, the authority was somewhat behind some of its fellow councils. Thus, as early as 1970, central government noted that: "computer systems in local government were developing very rapidly."[14] For example, West Sussex County Council had recently argued strongly for what amounted to a computerised population register and a common numbering system for the county. The council claimed that its personal reference number would allow it to: identify people from minimal information, locate anyone in its area, provide a means of collating information about groups of the population and streamline records by avoiding duplication. The county's version of a population register, the "personal master index," was to draw together information from all the authority's records of anyone the council had dealings with, but its ultimate aim was for the system to encompass everyone in the county.[15]

Under the Conservative governments after 1979, the relationship between central and local government became increasingly fractious and the OPCS was clearly allied to the centralising force of data, namely central government, and along with the Government Statistical Service (GSS), it more broadly had a history of ambivalence, if not outright opposition, to local data initiatives. In Chapter 3, it was shown how these tensions were part of the centralisation of politics considered under the Joint Approach to Social Policy. Thus, while tensions between different sites of power are inevitable in any political system, within Britain, these were exacerbated by the centralisation that was inherent in the growth of the data state. As James C. Scott has commented, the drive to create a thoroughly legible society, such as was being attempted here with the revised electoral register, "eliminates local monopolies of information" and, as a direct corollary of this, further creates "new positional advantages for those at the apex."[16] For example, at a 1972 meeting of the Central Statistical Office's (CSO) Committee for Statistics for Social Policy, earlier ambiguous phrasing was strongly criticised for seeming to suggest that records should be compiled and held at local Department of Health and Social Security (DHSS) offices, rather than within the central ministry.[17] This was because were records kept locally, they would be numbered and indexed locally and the different systems that would grow up around this autonomy would produce yet more islands in the data archipelago. Moser was in the chair for this meeting, and he recognised that, by the time he retired in August 1978, devolution was a rising political current in Britain. He maintained that the dispersed nature of the GSS would equip it to respond to any changes that this drift of constitutional policy might produce, but he insisted that the service should always, "come what may," aim to produce data for the population of the United Kingdom as a whole.[18] Such attitudes continued to colour the relationships between central and local government under the Thatcher/Major governments. Thus, a 1986 meeting of the CSO's Committee on Computing for the GSS noted approvingly that its new system allowed it to impose "more detailed monitoring," in order to better take "selective action" ("in this most politically sensitive area"), "against authorities whose spending was deemed to be excessive and unreasonable."[19] In this sense then, the British data state was built around what Nikolas Rose has called "fidelity techniques" for ensuring that the centre can maintain a hold over those on the periphery of the grid of power.[20]

The centralising function of the data turn in modern British politics is clear and it is clear that the central state desired to extend its gaze not only over the population, but also to encompass those other components of the state apparatus nominally beyond its control. In this manner, any accommodation between the local authorities that controlled the existing electoral registration scheme and central government could easily become at least somewhat fraught. The report on the proposed extension to the electoral register noted that local authorities were as hungry for data as the powers at the centre and were developing their own systems. They were also asking for an enhanced electoral register, but the report was adamant that, though localism could not be removed from the terrain of British politics, "it is important that these local initiatives should be co-ordinated in order to prevent inconsistent developments taking place in different areas."[21]

Given everything that has been seen about the centralisation involved in creating the data state and the impatience with which many of its leading lights grasped after more information about the population, this willingness to maintain the electoral register as a local survey may seem surprising. However, altering the electoral register so that it became a centralised document would involve complex legislation whereas increasing the scope of the register would be a relatively straightforward process.[22] Moreover, the report insisted that it was local authorities that needed this data, and they needed it to be available as rapidly as possible. It thus made good sense for the information to be both collected by, and to remain in the keeping of, these local bodies. Notwithstanding this apparent display of largesse towards local authorities, and though they were to be left charge of this project, they were not given a free rein on how they might go about it. Thus, the report sought to standardise the way this work was conducted so that the end product would be better than it currently was, making it easier for the centre to link together the various data tributes offered up to it from local sites. This may not have produced the perfect set of data that the OPCS was always looking for, but whatever it did produce should, it was suggested, be better than that forwarded by the current electoral registration system. Moreover, the new system held out the hope that given enough time, during which the centre could give more guidance to local authorities, it would eventually produce outcomes of the reliability and quality of the census.[23]

The previous chapter looked at the results of some of the exploratory research conducted for the original, mid-1970s, iteration of this proposal. This was followed up by two larger surveys. The earlier, exploratory, survey produced the hostile attitudes discussed in the previous chapter. However, the later survey did not produce anything like these comments and, to some extent at least, those conducting this second round seemed to be aware of this anomaly. This later research went through two versions, the first of which produced results, which even the OPCS advocates of an enhanced electoral register regarded as "artificial."[24] This first survey was conducted nationally, using standard sampling techniques across England, Wales and Scotland. It presented people, at 1,289 addresses, with one of three versions of the proposed new registration forms and asked them if they objected to what the forms stated would happen to their information. The first type of form was the standard registration form in use across the United Kingdom. The second stated that once the data on it had been used for statistical analysis, it would be destroyed. The third said that after statistical analysis, the form would be kept to allow the next year's survey to be completed more quickly by simply asking the respondent if anything had changed over the intervening twelve months. This survey found that only 2% of respondents had serious objections to being asked to complete any of the forms and most people would be happy were completion to be compulsory with very little difference across the subsets. Moreover, a majority of the people surveyed would be quite happy to have their completed forms kept if this would make it quicker and easier to complete the form in future years.

These results, which were too good to be true even for the advocates of the revised system, were the products of the nature of the survey. The original, exploratory, survey in 1977 had shown people a sample of the type of form that might have been used by the extended electoral canvas, explained how it differed to the status quo and recorded verbatim people's comments. However, the first of the two later surveys did not ask people if they minded increasing the flow of their personal information to the state since it seemed to assume that the public would remember what data they had previously provided when registering to vote. Moreover, it did not specify that they were actually being asked whether they would agree to make what had hitherto been a voluntary choice, to register to exercise their democratic rights, into a compulsory trawl of their personal data. Nevertheless, even this survey revealed that the people were less concerned about each particular piece of data that the government was suggesting they should provide than they were about the aggregate impact of this power/information shift. Even these "artificial" results thus chimed with public wariness that data provided to the government could lead to the production of dossiers.[25]

The second survey was conducted differently. It focused on three areas Warwickshire, Leicester and Bradford, where it piggy-backed on other local authority surveys, and its findings were less sanguine for the OPCS. In Bradford, 12% of respondents said that they would object very strongly to this revised registration process being compulsory, in Leicester, 9% had this view, while in Warwickshire, it was 8%. Taken as a whole, and compared with the first survey, this meant that the second survey showed that about twice as many people objected to the compulsion inherent in this proposal.[26] In addition to this revised benchmark for public attitudes, this survey came with three caveats. First, it admitted that it was impossible to assess how much opposition to a compulsory survey there might be were one really taken, because it was impossible to know in advance how much publicity this might attract, or how much of an impact this might have. Second, it was impossible to assess how accurately people might complete the canvas form. Here, it is worth remembering that these surveys were conducted around the same time as the Harringay trials for the 1981 census (see Chapter 4) where interviewers similarly reported that members of the public had entered their own personal details incorrectly. Third, the survey was based on interviews with people who had voluntarily responded to requests for their time and opinions, any future action/inaction of those who had refused to participate could only be guessed, but it was assumed that they would be hostile.[27]

Despite these caveats, the OPCS drew two clear conclusions from these surveys. First, that the public would be more ready to accept an extension to an existing survey that was already compulsory than to have either a new compulsory survey introduced, or to have what had always been a voluntary survey made compulsory.[28] Indeed, this was an argument that had been made before. Thus, in 1968, when the Wilson government was discussing its *People and Numbers* plan, it was suggested that the scheme could be made palatable, if not wholly acceptable, to the public if it were presented as an amalgamation of the electoral register and the census.[29] Second, the OPCS understood that in order to win the public over to its schemes, it would be necessary to engage them in a political dialogue. This emerged most clearly from the first survey where one of the versions of the forms shown to people stated clearly why the data was being collected.[30] It is worth reiterating at this point that though this research was conducted in 1977, it was published in 1981. In other words, it came after the Harringay protests had exposed the hollowness of any claims, which the OPCS or the Registrar General may have made, to be willing or able to engage the public in meaningful political dialogue. Yet despite this, and even in spite of the report's caveats, the OPCS still concluded that: "our work has suggested that the great majority of the public would accept a legal obligation to provide the extra information."[31]

In the end, this plan was abandoned. This was announced as an answer to a parliamentary question on 27 June 1980. This decision, it was stated, was based on a number of factors. The first of these mobilised the government's stock argument, for proving that it listened to what people said they wanted, by stating that government was concerned about the increased burden that an annual trawl might place on the public. Second, there was the problem of reconciling the traditional role of the electoral registration officers with the demands of the OPCS. Third, there was the recognition that the electoral register allowed some people, such as students and those with a second home, to register in two places, which would be incompatible with the requirements of the OPCS for accurate population data. Fourth, in this Rayner-driven political environment, there was the cost of all this.[32]

2

Margaret Thatcher's government introduced a reform to local government finances. This abolished the household rating system that had long been anathema to the Conservative Party and replaced it with the poll tax (officially called the "Community Charge"). This new system was first introduced in Scotland in 1989 before being rolled out across England and Wales a year later. The poll tax produced waves of popular protest across the country that eventually led to Thatcher's resignation and subsequent replacement by John Major as prime minister in late November 1990. The legislation to replace the poll tax was in place before the 1992 general election, which Major won, and the tax was abolished in 1993. The poll tax thus occupies an important place in the history of British popular protest and is, more generally, an important pivot point in modern British political history. However, its importance for the purposes of this book is the fact that this tax was to be levied and paid through a database known as the poll-tax register.

As part of the government's collection of ministerial views to coordinate a response to the Home Affairs Committee's suggestion of a study of common numbering, Michael Heseltine (by now in cabinet as secretary of state for the environment) wrote to the home secretary, Kenneth Baker. Heseltine noted that the mass non-payment of the poll tax was causing disquiet. He also stated that the audit commission had seen how one important element behind the success of this mass non-payment was that the "charging authorities have only limited scope for data exchange."[33] This meant that they could not locate defaulters who moved to another area even if these people registered for the tax in their new area. Heseltine agreed that this made a good case for Britain having "a common national identification number," or "a common PIN." Nevertheless, he was shrewd enough to add that the introduction of such a scheme purely to serve the poll tax was not something the government should consider. The scheme itself would already be "very controversial," so any association with the even more controversial poll tax would produce a truly toxic political brew.[34]

Here, in 1991, with the poll tax in its death throes, Heseltine put his finger on the essential weakness of the data that was supposed to underpin the tax. However, Philip Redfern had highlighted this essential flaw in the system as early as 1986 when he looked through the government's green paper on the poll tax.[35] The following year, during discussions about a proposed identity card scheme (see Chapter 7), it was suggested that poll tax collecting officers might find their jobs easier if they had "a national record of all UK residents," and also that a requirement to carry an ID card could similarly help local authorities to check that people had registered to pay the poll tax.[36] In 1988, Nicholas Ridley, the secretary of state for the environment, agreed that such an ID card system would aid in the administration of the poll tax, but only if the system were rigorously enforced, which would make it disproportionately expensive and even more controversial than it would have been anyway.[37] Redfern caught the essence of these points in an article published in the *Journal of the Royal Statistical Society* in 1989. Here, he stated the real problem was that the poll-tax registers were to be stand-alone data sets: data islands. Thus the law did not make any "provision ... for relating the CC [Community Charge] registers to the most comprehensive of the national registers, the NHSCR [the National Health Service Central Register], or to the extensive set of registers held for social security purposes," all of which was compounded by the fact that even if such provision had been permitted, it could not have been realised as the poll-tax registers listed people only by name. These shortcomings were, he argued, utterly lamentable. But more importantly from his perspective, they could have been prevented by the use of common numbering or population registration. But these courses of action had not been followed and so he concluded that: "the quality of being uncoordinated and therefore error-prone is one of the design criteria of the new system."[38] The poll tax was thus destined to fail.

That this analysis was both obvious to Redfern in advance of the actual events that led to Heseltine's comments, and equally obvious to Ridley as events unfolded must beg the question: why was the system built around such a gapping design flaw in the first place? The government had considered the whole issue of common numbering both publically and privately through the 1980s and it would continue to do so under John Major. It would also, as will be seen in subsequent chapters, pursue ID cards and computerised data systems. Nevertheless, the government itself was a shotgun marriage of modernising biopolitical imperatives with some deeply conservative political instincts and it should not be surprising that, even without factoring in the role of political contingencies, these two conflicted. For example, less than a year after she first took office as prime minister, Thatcher "expressed the strong personal view" that the original copies of census forms completed by the people should never be released for public examination. She argued that once these had been used for appropriate population data purposes, they should in fact be destroyed. The law stated that the forms could be released after hundred years, but Thatcher felt that this would weaken the pledge of confidentiality given to the public when they completed their returns.[39] On this occasion, the voices of those, such as Redfern, from within the GSS, orchestrated by the Registrar General, Roger Thatcher, prevailed. They argued that the census data was essential for historical and genealogical research and the longitudinal study, but more importantly for medical research (from which quarter a "strong reaction" to the prime minister's views could be expected).[40] Additionally, they added that since they had dropped the proposed questions on nationality and ethnicity, and had also proven that Thorpe's concerns (see Chapter 5) over confidentiality were much ado about nothing, the public had nothing to fear from the forms being retained in the usual way.

This brief episode might have all the appearances of a storm in a Whitehall teacup. But from the perspective of those ardent data enthusiasts inside the GSS, the prime minister's suggestion must have seemed utterly inconceivable yet also typical of ministerial political reflexes. Certainly Redfern seemed to detect a similar set of attitudes in relation to the government's design for the poll-tax system. After his retirement from the OPCS, in 1982, Redfern undertook a special study, for the Statistical Office of the European Communities, into administrative methods of conducting population censuses. He was, in other words, examining how the population could be surveyed by using administrative data in lieu of the type of overt surveying of the people that had produced opposition across Europe. Thus, he was an acknowledged expert on the subject and as such was something of a thorn in the side of his former colleagues. In his view, the root of the problem for the poll tax lay in the government's lacking the courage of its own biopolitical convictions. Making this point, he quoted the government green paper as stating that while in other countries, the authorities had unified their administrative data into a central registration system: "the British tradition on registration is different. Registers are kept separately for different purposes ... There will be no national register."[41] The government had opted for a shotgun wedding with the traditions of localism inherent in Britishness, rather than apply what Redfern argued was "rational thought" to design and implement a system that was tailor-made for the job in hand and, in this instance, the wedding had produced a calamitously bad union.[42]

3

When the data protection registrar, Eric Howe, appeared before the House of Commons Home Affairs Select Committee in December 1990, there was a discussion about how credit reference agencies wanted to use personal identification numbers for each person listed in their databanks. This discussion subsequently encompassed the use of personal identification numbers more broadly across government. As part of this discussion, Howe expressed his concern that the creeping use of an existing system, such as that used by the national insurance (NI) system, could lead to the creation of a common numbering system and, on this basis, argued that the use of any numbering system should be strictly limited to the system for which it had been originally designed.

However, the committee had before them a paper written by Philip Redfern.[43] This was the paper (see Chapter 4) in which Redfern made the case for the advantages of the panopticon-style properties of population registration, and (see Chapter 5) where he lambasted the hesitancy of government in facing up to the opponents of the data turn. The committee quoted one of Redfern's stock phrases in describing the government's data system as "higgledy-piggledy" and followed him to argue that it was "apparent that the present means of identification are ill defined and prone to inaccuracy and abuse."[44] The committee believed that the country clearly lacked an efficient system of identification and accordingly concluded that the Home Office "should initiate a comprehensive study … of the case for and against a common national identification number."[45]

As has been seen, Redfern had actually retired from his position as deputy director of the OPCS in 1982, but he kept abreast of developments in both his field and the corridors of Whitehall. The select committee's report was published before Christmas 1990 and by the middle of the next month, he had already written to the registrar general, Peter Wormald, to ask to meet him to discuss the report's recommendation. Moreover, Redfern was not alone in coming back to Whitehall from beyond retirement to lobby on this issue. Roger Thatcher, one of Wormald's predecessors, was also a keen advocate of population registration and was applying similar pressure to Wormald. Elsewhere along Whitehall, Sir Jack Hibbert, the chief statistical officer, was contacted by one of his predecessors, Sir John Boreham, who made the same point.[46] Wormald wrote to the Home Office requesting that the government agree a line to take on this recommendation by the select committee. He argued that though he was keeping Redfern at bay, he could not do so forever as they met in the ordinary course of their business and he was concerned that if it became public knowledge that he was talking to one of the known proponents of population registration, then this might be seen as the government's acceding to the select committee's conclusions.

The initial response to Wormald's pleas for some clarity with which he could repel Redfern came from the home secretary, Kenneth Baker. This produced the form of words that was eventually used in the government's reply in March 1991. This was that the government had "no present intention of commissioning detailed work on such a proposal."[47] This may have made its way into the final paper, but it did not do so without attracting criticism. Baker wrote that the topic was "exceedingly sensitive politically, since it will be seen as opening up the possibility of comprehensive tracking of a citizen's movements and activities."[48] Most of his cabinet colleagues agreed with this wariness around public sensibilities and, as a result, many were concerned that his use of the phrase "no present intention" might seem to indicate that such a policy initiative was in the pipeline for the future. For example, William Waldergrave, the health minister, wrote that he was concerned that even "a hint that we might introduce" such a scheme could be politically damaging, while Eric Thompson, Wormald's deputy, was concerned that this glimpse of a crack in the government's position could "prolong the pressure from Mr Redfern and others."[49] In other words, these officials and ministers were concerned that Baker's rejection of common numbering was not robust enough.

In March 1991, the government published its reply to the Home Affairs Committee report. This response tartly pointed out that the committee had considerably exceeded its brief by making this recommendation about common numbering and, using Baker's phrasing, kicked the committee's proposal for a comprehensive study of the subject into the long grass. Given what has been seen of this government's attachment to population data, along with its eagerness to gather more in its increasingly centralised institutions, this reluctance to embrace common numbering might seem contradictory. However, the reality of the situation was that this public dismissal of the idea of common numbering followed the pattern set by those officials, interviewed by the Lindop Committee, who had issued assurances that the government had no intentions of bringing a common numbering system into play. As on that earlier occasion, here in 1990–91, this declaration concealed more than it illuminated.

At this point, early 1991, Peter Lilley was secretary of state for trade and industry and as such he had oversight of credit reference agencies. He wrote, in reaction to Baker's suggested response to the committee, to support the home secretary's form of words. However, this was not because he opposed data linking and the use of common numbering. Indeed, in 1992, he would enter the Department for Social Security, bringing with him an anti-fraud zeal that would be galvanised by data linkage (see Chapter 4). More importantly, at this point, he was, as he told Baker, keen to see the national credit register developed and was aware that this depended on "access to comprehensive and sound information."[50] Similarly, Waldergrave's awareness of the possible political fallout from being seen as too close to the committee's conclusion was not designed to protect the people from data harvesting. He based his critique of Baker's apparently ambiguous language on the fact that any opposition this might stir up could rebound on the census that was due to be taken that April. The census was, obviously, administered by the OPCS for which, as health minister, Waldergrave had responsibility.[51]

Understandably, the OPCS shared Waldergrave's concern to protect the census. Wormald noted that already, in January, opposition to the census seemed to be gaining momentum building on public fears that it might be linked to the poll-tax registers (see above). He argued that the civil-liberties issues raised by the committee's reports were beyond the scope of his office, but wanted people across government to note that:

> we rely on the voluntary cooperation of the public to acquire a lot of information which is indispensible to the service which we provide ... Any move which threatened that cooperation ... could have the most serious consequences for OPCS statistics and for the Social Survey.[52]

More importantly, Wormald argued that if all this damage, caused by protests against common numbering's potential to usher in de-facto population registration, were to ensue, it would all be for nothing because "this country already has an almost complete register of the population."[53]

This was the NHSCR. All children born in the United Kingdom were given an NHS number when their births were registered and virtually everyone had, at some stage, been issued with a number-bearing NHS card. However, Wormald conceded that the NHSCR was not perfect. If people moved without reregistering with a doctor at their new address, then their data would be out of date, while newly arrived immigrants were similarly often absent from its lists. These shortcomings were, for data ultras such as Redfern, enough to damn the NHSCR's ability to ever function as a true population register. But as Wormald saw this as a difference of degree, not kind. Moreover, the Department of Health was actively considering using the NHSCR more widely and the OPCS itself was very interested in this "because wider use within the NHS and [the] OPCS could substantially enhance the scope and quality of the health statistics."[54]

Moreover, within two years, the registrar general's office would write that though "the idea of a general purpose register is not wholly accepted ... personal identification systems and holding and linking of personal information in data banks are becoming more commonplace."[55] These trawls, it argued, were set to continue, despite any privatisation programmes the government might enact, as the use of data within government, and the demands for it from the EC were relentless, as were demands for cost-cutting imposed across the GSS. These were the points made by Bill McLennan when he became chief statistical officer in 1992 and would underpin his drastic overhaul of the government's data architecture (see Chapter 5).

Increasing the use of data already held by government, through linking it across the data silos of different departments, or across systems within a department, was not something new in the early 1990s. Indeed, Foucault has shown how the "considerable extension of procedures of control, constraint and coercion" inherent in this data-based panopticonism were "the counterpart and counter-weight of different freedoms" and that "liberalism ... and disciplinary techniques are completely bound up with each other."[56] Thus, when Acheson proposed that his common numbering scheme could be extended to any data system, ultimately linking all such information repositories, he was, as Foucault points out, echoing Bentham who proposed that:

"the Panopticon should be the formula for the whole of government saying that the Panopticon is the very formula of liberal government."[57]

However, in the shorter term, as was seen in Chapter 4, the Rayner review of 1980 boosted these tendencies. For example, in 1984, Norman Fowler, minister at the DHSS, received word from his officials that under Peter Brooke, a Treasury minister, the government was moving ahead with shared computer-data networking. The officials first had to didactically outline to Fowler what a computer network actually was, which they achieved by comparing it to the BT phone network. Then, once this had been sketched out, they moved on to wholly endorse the suggestion. Officials from the DHSS, the Home Office and the Inland Revenue had recently visited the United States where they had seen data sharing in operation and had returned home inspired and fired up with zeal for such developments.

The DHSS, these officials told Fowler, already had its own network of 6,000 terminals that was being boosted by a national rollout to all unemployment benefit offices. Moreover, and indicative of the trend, these officials told the minister that the DHSS had

> for some time been suggesting to the Inland Revenue that the tasks of the two departments and the way in which offices were spread across the country meant that our requirements were in technical terms remarkably similar and we should consider a joint network.[58]

After which they added that such a joint network could easily become the backbone for one used across all departments with the exception of the Ministry of Defence, which had its own particular needs. Here they suggested that the Post Office and the NHS would be prime contenders to join such a network but stressed that this type of project could "provide the infra-structure for government communication into the very long term … improving the whole business of government since the accurate and efficient communication of information is one of the keys to sound management."[59] When this system was developed, it was called the Government Data Network and is discussed more fully in Chapter 8.

These officials were aware that this plan would concern the public, concern that would only increase once the Home Office's interest in the scheme became known: "images of '1984' and a 'Big Brother' network," "some sinister tool of the state which will allow the police to browse freely through Social Security and tax records" were, they dismissively added, "thrown about with vigour."[60] But, they argued, these were issues to be managed; they were not reasons to modify, never mind abandon the scheme and the way forward in handling the public was clearly signposted by the nature of the system. This was to be a computer-driven network and what the public needed to be told was that: "the technical ability exists to keep all access to information under rather tighter control than is possible in a paper system and the network … can be designed to only allow authorised access."[61] The veracity of these claims is assessed in Chapter 9, but, following the general pattern of government utterances about its data gathering, it is clear that in lieu of any political discussion about the nature of the information pooled, or why this information was held in government files at all, the public would only be offered technical reassurances. Thus, as the system was rolled out, the depoliticisation that came in its wake rolled on.

These plans to link data sets depended on the sets being able to connect information they held about a given individual. Doing so by using names was, as has been seen, at best extremely difficult since the usage of names changed so much.[62] All of which meant that linkage needed some form of common numbering. There were two such numbering systems in use that were usually suggested as candidates for this type of wider use. These were the NI and NHS numbers. NI numbers, however, were only given to people who needed them. Thus, for example, children were beyond their reach, which left the NHS numbers as the prime target for those who wished to broaden the use of an extant system of numerical identifiers. Within the DHSS, this was discussed in 1986 when proposals were mooted to use the NHS numbers to link all the department's systems together. However, this was rejected for two reasons. First, because doing so would have been unacceptable to the medical profession. Second, because the NHSCR, although called "national," was compiled and maintained locally and any extension of the use of its numbering system would have entailed more centralisation that would, in turn, have involved huge upheavals to the whole health system.[63] However, at this time, the department's various data sets were being computerised and the new systems needed numbers to identify people. Given that numbers were going to be assigned for this purpose, it made sense, from this perspective, to assign one number to each person the department dealt with.[64] This was certainly, and perhaps predictably, the view of the Management and Efficiency Division of the Cabinet Office who saw all this as a chance to gain yet more savings in staffing costs from computerisation than would otherwise have been the case.[65]

A briefing paper for a meeting to review this issue filled in the Whitehall history behind these plans. A management review of the working of the OPCS revealed that though the office was responsible for the maintenance of the records of births, marriages and deaths, it actually had very little control over the policies that governed the data that formed the basis of this work. Marriages, for example, were governed by policy made in the Home Office, which the OPCS from its Health Department base was obliged to follow. The upshot of these investigations and discussions was the conclusion that all the departments interested in this data would find their interests better served "by the introduction of a common identifying number for each citizen resident in the UK. … This, it was conjectured, would lead to far greater cohesion, efficiency and effectiveness."[66] The paper then listed five advantages that would flow from such a system. Three of these focused on combatting fraud, one mentioned savings in staffing costs and only one centred on an improved service to the people the DHSS existed to serve. The paper further noted that even if the mergers of NHS and DHSS numbers did not go ahead, the DHSS was definitely going to push ahead with its own scheme to address similar issues. However, it was also noted that there would be opposition to this plan. This, it was said, would be built on fears that ran along a spectrum that included the belief that the change could herald a move towards a much more totalitarian form of government. While this fear was "perhaps not wholly justified," the government, it was suggested, might flinch from taking on the "powerful lobbies" that propagated such ideas.[67]

In the end, the government was not required to face up to any opponent over this issue as it dawned on those involved that the scheme had been rendered impracticable. This was because the registrar general had already introduced a new NHS numbering system to facilitate NHS computerisation. This had been accepted across the health system, and this guaranteed that it could not be altered or its application modified.[68]

The explanation most often given for data-linkage systems organised around common numbering was that they were needed to provide information for statistical purposes. Philip Redfern was a statistician who spent most of his working life at the heart of the government's data operations, but he did not accept that the justification for any of the systems he devised or steered could be found primarily in the provision of statistics. In fact, he argued that the real purpose of these systems "rests on consideration of administration and policy," "to enable governments to exercise a wider range of policy options."[69] The statistical issues were thus, as he noted elsewhere, "important but they are secondary."[70] On this point, it is worth noting the comments made by the government's own National Information Infrastructure Taskforce in 1997, a point when the government was considering introducing a computerised interface system, called *government.direct*, for all interactions between its service providers and the public (see Chapter 8). Here, despite the technical nature of both the project, and the body commentating, the conclusion reached was that: "clearly the availability of technology is not a fundamental issue, it is the extent to which ... governments drive the applications which achieve their end objectives."[71] As Foucault noted (see Chapter 1), the attitude and political determination of the government was the driving force of these "applications" designed to push government's "end objectives."

As it happens, Redfern not only championed the government's applications, to the extent of chivvying official timidity over their implementation, he also supported the government's political objectives. When he listed reasons why the government needed to improve its data systems, the factors he gave were: to prevent fraud in the social security system, to control immigration, to introduce a poll tax and to overhaul the electoral registration system in order to prevent voter fraud.[72] This agreement over the end objectives is, of course, important, but more fundamental was the fact that Redfern could write to the director of operational strategy at the DHSS to discuss the applications designed to achieve what he termed: "the quality of society that we seek."[73] He was exasperated that ever since he had joined the OPCS in 1970, government had failed to inaugurate the panopticon-style applications it needed in order to achieve its own ends. From this point of view, the problem with the poll-tax registers was that they were the product of political reflexes. They had been formed by kneejerk responses to perceptions of a hostile public because the government lacked the courage of its own convictions.[74] However, as has been seen, behind the scenes, the systems for linking data sets through common numbering were, despite public denials to the contrary, moving ahead. Indeed the next two chapters will detail developments, along these lines, which would increasingly become the new normal of government data gathering.

Notes

1 OPCS Population Statistics Division, *Report of the Steering Committee to the Registrars General, Extending the Electoral Register – 1* (London: OPCS Occasional Paper, 20, 1981), 32.
2 Ibid., 25 and 2.
3 Ibid., 38.
4 Ibid., 37 and 33.
5 Ibid., 35.
6 Ibid., 35.
7 Ibid., 26–28.
8 Ibid., 39.
9 Ibid., 5.
10 TNA, HLG 156/787/2, West Yorkshire Metropolitan County Council, Economic and Commercial Policy Group, *Information and Data Collection and Usage*, n.d., ca. 1979, 2.
11 Ibid., 4; TNA, HLG 156/787/2, South Yorkshire County Council, Data Sharing Working Party, 19 Jan. 1979, 1–10; and TNA HLG 156/787/1, Joint Working Group of the Yorkshire and Humberside Regional Board and the Strategic Conference of County Councils, *Data Sharing, Employment and Industry*, Apr. 1980, 2–8.
12 TNA, PREM 16/1884, Sir Claus Moser, *Review of the Government Statistical Service*, 4 Aug. 1978, 30.
13 TNA, HLG 156/787/2, South Yorkshire CC, *Soft Information Systems*, 2.
14 TNA, HO 411/1, *Code of Conduct for Government Computer Systems, Meeting*, 4 Mar. 1970, 3.
15 TNA, CAB 139/563, West Sussex County Council, Computer Steering Group, *Report of the Working Party on Personal and Locational References*, n.d., ca., 1970, 5.
16 James C. Scott, *Seeing Like a State: How Certain Schemes to Improve the Human Condition Have Failed* (London: Veritas, 2020), 78.
17 TNA, BN 89/202, CSO, Committee on Statistics for Social Policy, *Meeting*, 9 Nov. 1972, 6.
18 TNA, PREM 16/1884, Sir Claus Moser, *Review of the GSS*, 55.
19 TNA, CAB 108/424, CSO, Committee on Computing for the Government Statistical Service, *Report on GSS Computing During 1985/86 from CSO to the Head of the GSS*, 30 May 1986, 14.

20 Nikolas Rose, *Powers of Freedom: Reframing Political Thought* (Cambridge: Cambridge University Press, 1999), 212.
21 OPCS Population Statistics Division, *Extending the Electoral Register – 1*, 5.
22 Ibid., 42.
23 Ibid., 15.
24 OPCS, *Extending the Electoral Register -2: Two Surveys of Public Acceptability* (OPCS: Occasional Paper, 1981), 21.
25 Ibid., 8.
26 Ibid., 24.
27 Ibid., 35.
28 Ibid., 30.
29 TNA, HO 328/130, Children's Department, *A Population Register and Common Numbering System for Great Britain*, 15 Oct. 1968, 1.
30 OPCS, *Extending the Electoral Register – 2*, 17.
31 OPCS Population Statistics Division, *Extending the Electoral Register – 1*, 18.
32 Ibid., Insertion between pages 24 and 25.
33 TNA, PREM 19/3379, Heseltine to Baker, 8 Mar. 1991, 1.
34 Ibid.
35 TNA, RG 22/54, Redfern to Spackman, 3 Dec. 1986, 1.
36 TNA, HO 301/314, *National Identity Cards*, n.d., ca., Feb. 1987, 4.
37 TNA, LAB 109/88, Ridley to Hurd, 14 Oct. 1988, 2.
38 Philip Redfern, "Population Registers: Some Administrative and Statistical Pros and Cons," *Journal of the Royal Statistical Society, Series A*, 152 (1989): 19.
39 TNA, RG 19/778, *1981 Census: Preservation or Destruction of the Census Forms after Analysis*, 23 Jan. 1980, 1.
40 TNA, RG 19/864, *Preservation or Destruction of the 1981 Census Returns*, 1 Feb. 1980, 1.
41 Quoted in, TNA RG 22/54, Redfern to Spackman, 3 Dec. 1986, 2.
42 Ibid.
43 Philip Redfern, "A Population Register or Identity Cards for 1992?," *Public Administration* 68 (1990).
44 Home Affairs Committee, Session 1990–91, *First Report, Annual Report of the Data Protection Registrar*, 12 Dec. 1990, x.
45 Ibid.
46 TNA, RG 50/23, Wormald to Langdon, 22 Jan. 1991, 2; and, Boreham to Hibbert, 22 Jan. 1991, 1.
47 TNA, PREM 19/3379, Baker to Waddington, 6 Feb. 1991, 2.
48 Ibid.
49 TNA, RG 50/23, Thompson to Wormald, 12 Feb. 1991, 1.
50 TNA, PREM 19/3379, Lilley to Baker, 21 Feb. 1991, 1.
51 TNA, PREM 19/3379, Waldergrave to Baker, 26 Feb. 1991.
52 TNA, RG 50/23, P. J. Wormald, *A Common National Identification Number: Home Secretary's Letter of 6 February*, 6 Feb. 1991, 2.
53 Ibid., 1.
54 Ibid.
55 TNA, RG 50/34, *Statistical Information on Population and Housing (1996–2016)*, 10 Feb. 1993, 8.
56 Michel Foucault, *The Birth of Biopolitics* (New York: Palgrave MacMillan, 2010), 67.
57 Ibid.
58 TNA, HO 524/3, *Shared Data Networks*, Memo with Spackman to Laurence, 4 July 1984, 1.
59 Ibid., 2.

60 Ibid., 3.
61 Ibid.
62 TNA, HO 524/1, *IND Computer Strategy Meeting*, 12 July 1982, 3.
63 TNA, RG 22/54, Preston to McCulloch, 17 Apr. 1986, 1.
64 TNA, RG 22/54, Jordan to Alder, 15 June 1986, 1.
65 TNA, RG 22/54, Pennington to Philips, 29 Oct. 1986, 1.
66 TNA, RG 22/54, *Unification of DHSS and NHSCR Identity Numbers*, 14 Nov. 1986, 2.
67 Ibid., 3.
68 TNA, RG 22/54, Dainty to Brown, 26 Nov. 1986, 2.
69 TNA RG 22/54, Redfern to Spackman, 3 Dec. 1986, 1; and TNA, RG 22/54, Redfern, *Which Countries Will Follow the Scandinavian Lead in Taking a Register-Based Census of Population?*, n.d., ca. 1986, 1.
70 Philip Redfern, "Population Registers: Pros and Cons", 2.
71 TNA, CAB 202/135, National Information Infrastructure Taskforce, *Response to the Green Paper 'government.direct'*, 4 Feb. 1997, 2.
72 TNA, RG 22/54, Redfern, *Which Countries Will Follow the Scandinavian Lead?*, 15.
73 TNA RG 22/54, Redfern to Spackman, 3 Dec. 1986, 2.
74 Ibid.

7 Data Systems 1979–97 (2), Driving Licences and ID Cards

It is obviously the case that states have always looked at their societies but as James C Scott has pointed out, the degree to which the society is legible to the state's gaze has changed markedly and, as the society has been opened up, so the range of activities that the state can pursue has increased. However, the change from the earlier opacity, where the state's interventions were "largely confined to grabbing a few tons of grain and rounding up a few conscripts," to modern legibility where "the state's objective requires changing the daily habits … or work performance … of its citizens" does not occur on political terra incognito.[1] In fact, it is enacted on a political landscape that has been both mapped and shaped by the previous system. As a result, the transformations are applied to a population who have their own customary cultural and political assumptions formed through relationships originating in this system. In the situation examined here, these were the assumptions and attributes of British political culture. Thus, as was discussed in Chapter 2, and as Claus Moser found to his chagrin, data systems for the new range of modern biopolitical interventions can rarely be drawn de novo and so their advocates are often obliged to adapt what is already at hand. This is what was attempted with the OPCS's plans for a revised electoral register and it also characterised the attempts to extend the use of National Health Service (NHS) or National Insurance (NI) numbers to form a common numbering system.

The first section of this chapter looks at another attempt to broaden and deepen the use of an extant data system: driving licences. These are important not only because of their very prosaic nature, indeed apart from birth certificates, they are still perhaps the government document most commonly held by British people, but also because the plans that were spun around these licences reveal a lot about the impetus behind the government's push for population data. The plans developed by these Conservative governments, centred on a new photo-bearing format for UK-wide driving licences, snowballed and formed the basis of attempts to introduce national ID cards. This country-wide attempted reform was explicitly built on the example of the licences already in use in Northern Ireland, and the situation in this corner of the United Kingdom is examined in the second section

DOI: 10.4324/9781003252504-8

of this chapter. The third section analyses how this reform to the driving licence, and the ensuing discussion of how the licence could become an ID card, escalated yet further with suggestions that, were a UK ID card turned into a smart card, it could be used to access services in both the public and private sectors. Indeed, as this function creep set in, the potential involvement of the private sector increased, revealing the extent to which these plans were rooted in government perceptions of, and its involvement in, the increasing commercialisation of data.

1

A form of registration for drivers was first introduced in Britain in 1903. The general permission to drive granted by this system was subject to later restrictions governing the type of vehicle the holder could drive, with driving tests being introduced in 1935. This system, barring a few changes to the licence itself, remained in situ until the mid-1960s when the system was centralised. Before 1965, driving licences (and the paper tax discs displayed on all road vehicles) were issued by local authorities, but, at that point, the system was placed in the hands of the Driver Vehicle Licencing Centre (DVLC), a body that was restyled as the Driver Vehicle Licencing Agency (DVLA) in 1990. That this mid-1960s process of centralisation was all of a piece with the developments discussed in previous chapters, is made clear by the fact that these processes went hand-in-hand with "discussions of using either the NHS or National Insurance or a combined number" to facilitate this centralisation.[2]

Driving licences not only validate a person's competence as a driver, they also identify the holder and, in 1986, it was this feature of licences that was uppermost in the minds of the Metropolitan Police. That year the police gave evidence to a Road Traffic Law Review (the North Review) and the discussions at this review drifted on to consider several issues outside the review's immediate terms of reference. The force circulated these issues to its branches, collated their responses and sent them as a supplementary report to the review. This began by presenting the anomalous situation whereby motorists stopped by the police, who did not produce identification, or who gave false particulars, could evade responsibility for an offence, whereas those who identified themselves might well have to face the consequences of their actions. Given this situation, the force reported that its suggestion that there should be a statutory requirement for motorists to carry a driving licence, had received unanimous support across the branches surveyed.[3] The police also had opinions on what the licence should look like. Licences in Northern Ireland had recently been reformatted to include a photo of the holder. This needed to be updated every ten years and this provided a means of ensuring that licence holders updated their addresses reasonably regularly too. A proposal to adopt this type of licence across the rest of the United Kingdom received the fulsome support of the Metropolitan Police.

However, the force also proffered a compromise solution. This plan would not have made a failure to carry a licence a criminal offence, but it would have given the police the power to detain anyone driving while not in possession of a licence, until such time as they had been satisfactorily identified. Throughout this paper, the force was keen to emphasise that none of these changes it advocated would represent the introduction of an identity card in Britain since "no one is forced to drive."[4] Holding a driving licence was, the police insisted, a personal choice, not a legal requirement.

The initial response of the Police Department in the Home Office to this was to suggest that the North Review could be "steered" in the direction indicated by the police, but shortly after this, a different, and strikingly liberal counsel emerged within the department.[5] This pointed out that: "there is something rather objectionable about the creation of an offence which does not relate to the licence holder's care or competence ... and which would exist essentially for the convenience of the police." Moreover, it was argued that obliging people to carry licences whenever they drove would mean that more documentation would be left in cars and "what better gift to the car thief?" But most importantly this argued that the compromise brought forward by the police was in fact nothing of the sort. Rather the ability to detain people for not having a licence was a "draconian" power. Moreover, the official making this case could not see how the public would be able to distinguish this "detention" from being arrested. As a result, the point was forcefully made that any of these proposals would damage the relations between the public and the police but, as the author acerbically concluded, "perhaps relations with the motoring public are so poor anyway that the police don't mind."[6]

However, the original official from the Police Department returned to the fray early the following year (1987) by which time he had outflanked his opponent by undertaking "wide consultations within the Home Office." These extended the topic under discussion from the format and use of the driving licence to consider these alongside the possibility of introducing identity cards. This set of proposals began by suggesting that it was largely pointless to review objections to ID cards on the grounds of "so-called civil liberties," which the author damned with faint praise by conceding that they possessed "much emotional force."[7] However, he argued that even this force would decrease as Britain became more of a "credit-card society," where people became, like citizens of the United States, increasingly accustomed to carrying, showing and using plastic cards. He showed how, in the United States, driving licences were used as ID cards and noted that, in California, the licence-issuing authorities issued a version of the licence to non-drivers to allow them to identify themselves properly and so have legitimate access to "the pleasures of adult life." The US driving licence/ID card was thus: "a convenience rather than an imposition." On this basis, the author hoped that the North Review would stimulate discussion of the topic of identification documentation "on favourable grounds." However, he conceded that the facts that the public were still so steeped in the "British traditional distaste for documents" and that Britain was not yet a "credit card society" to the same extent as the United States, militated against the introduction of an ID card, which he frankly admitted "was not a runner." Nevertheless, he raised the issue of the format of the licence in use in Northern Ireland and suggested that whether or not licence carrying became compulsory, the North Review might like to discuss extending the use of this type of photograph-bearing document to the rest of the United Kingdom.[8]

These suggestions were passed up to the home secretary, who accepted that the arguments against identity cards, though often "overstated" were, nonetheless "strongly felt" and, on this basis, declined to commit the government to study the topic and decided to leave the North Review to study what it felt best.[9] The government's proposals in response to the North Review were published just less than two years later, in February 1989, and signed off by the home secretary and the secretaries of state for transport and Scotland: Douglas Hurd (he held this post from September 1985 until October 1989 when he became foreign secretary, see below), Paul Channon and Malcolm Rifkind respectively. These gave the police some of things they had sought at the review. The government would change the law to allow the use of speed cameras. More importantly, it committed the government to undertake a feasibility study to identify the pros and cons of increasing the linkage between the records of the DVLC and those held on the Police National Computer. However, it rejected the demands to make carrying a licence mandatory.[10] This may make it sound like the cause of reformatting the UK-wide driving licence, never mind that of a national ID card, was abandoned at the end of the 1980s, or that the police and their Home Office allies suffered a defeat. However, two things need to be noted here. First, that the increased access the police had to the data bases of the DVLC more than compensated for the failure to secure a compulsion to carry a licence. After all, even the force's advocate in the Home Office admitted that: "we have yet to hear complaints that the name and address provision [of the Police and Criminal Evidence Act] is being significantly undermined by people giving false names and addresses."[11] Second, this intervention into the discussion about road traffic law both opened up the discussion of extending the Northern Ireland format of the licence to the rest of the United Kingdom, and also provided some of the key arguments that would reverberate through subsequent discussions of driving licences and ID cards.

By 1991, Britain seems to have become more of a credit card society, because in that year, a consultation exercise for the Department of Transport revealed strong public support for a photograph-bearing, plastic, credit card-sized driving licence, with 72% of drivers interviewed preferring this to the existing paper licence.[12] Such a licence, it was stated, would be more durable, secure and convenient and, above all, would allow for the licence to be significantly upgraded to a smart card by the addition of a microchip (though it should be noted that the poll had not sought the public's opinion on this). This licence would be based on the experience of that used in Northern Ireland and as such would require the photo to be updated every ten years thus necessitating a periodic update of the holder's address. The photo used would need to be verified to prove it really was a likeness of the licence holder and, at this stage, a system for doing this was being canvassed. Those proposing the system wanted it to be based on a greater linkage of police and DVLA data which, it was hoped, would be achieved by the use of an electronic data interchange enabling police at the roadside to have access to any endorsements currently held by the licence holder. However, if technological shortcomings were to prevent this, they proposed a half-way-house, whereby endorsements would still be shown on a paper counterpart to the new plastic licence. Since it was recognised that this licence would rapidly become a proxy ID card, these proposals sought views on whether, as was the case in California, some form of voluntary card could be made available to non-drivers.[13] Though at this point it was felt that the scheme could be ready for introduction across the United Kingdom in mid-1996, these new licences were eventually introduced in 1998 and when they were, they came with a paper counterpart.

In his report of June 1991, the year the Department of Transport published its finding on public support for the new licence proposals, the data protection registrar expressed considerable concern about these developments. He suggested that there should be serious political consultations with the public before the police were given access to the DVLA data base and, notwithstanding ministerial reassurances to the contrary, that the new licence would become a de facto ID card, even if it were not to become an actual ID card. As events unfolded, these fears would prove to be substantial underestimations of the government's plans. The way these plans developed from an upgrade to the humdrum driving licence and snowballed into plans for a national smart ID card is the subject of the rest of this chapter with the next section focusing on where these moves began: Northern Ireland.

2

For much of the period covered by this book, Northern Ireland was in a state of armed conflict and, given this political-military situation, it is not surprising that it was there that debates around knowing/identifying members of the population were intensified within government circles. As has been seen, people in this part of the United Kingdom had photo-bearing driving licences before those in the rest of the country and it was here that the debates on national ID cards originated.

As far as the authorities in Northern Ireland were concerned "the determining factor in deciding whether or not to introduce identity cards ... [was] the overriding aim of greater security."[14] The British army had always opposed the piecemeal introduction of identity cards. It had always maintained that an identity card would only be of any real use if it covered not only the whole of the United Kingdom but the Republic of Ireland as well. This was why it opposed the introduction of ID cards that were geographically limited to Northern Ireland. Moreover, the army also objected to a card that might be designed to be used more widely, but only to be taken up by people who wanted to have a card: a voluntary ID card system. However, by late 1988, the military had realigned its thinking on ID cards to join the Royal Ulster Constabulary (RUC) in accepting that a province-wide and voluntary scheme would have some advantages. This was because the military realised that even if only law-abiding people carried ID cards, these would still speed up the processing of the estimated 1,200 identity checks carried out daily at the army's vehicle checkpoints and so would expose soldiers to less danger.

The system envisaged was designed expressly to piggyback on the Northern-Irish photo-bearing driving licence that was widely used as an identification document across the province. There were 800,000 of these licences in circulation, and 50,000 new ones were issued each year. The plan was that, from the point at which the new system came into force, an application for a licence would be treated automatically as entailing an application for an identity card, as would licence renewals. A California-style, non-driver's version of the card was also included in the plans to boost its uptake. The public would also be enticed into the system by the fact that these identity cards were touted as also possibly acting as cheque guarantee cards (used for ordinary banking purposes), a proof-of-age card, proof of identity for social-security purposes and (were the Foreign Office to agree) as a travel document. If these carrots of convenience did not work, the authorities were ready to apply a stick of coercion to drive people into the system.

Quite how far they were prepared to go with this, surely indicative of how much they wanted the system introduced, was made clear by discussions surrounding the anticipated opposition to the scheme. Here the government concluded that the two most promising methods of obliging uptake of the scheme were "the suspension or withdrawal of entitlement to social-security benefits; and making the right to vote conditional on the presentation of an identity card."[15] With regard to the former, while it was widely assumed within the security establishment that most paramilitary personnel were claiming social-security benefits, it was also recognised that the existence of an identity card could not ensure that officials in remote offices, perhaps subject to paramilitary pressure, would actually ask to see a card before paying out money. Moreover, it was also conceded that, within the social-security system itself, it was understood that most fraud could not be stopped by use of an identity card (see Chapter 4).[16] But the money was only part of the point at issue here. As the discussion made clear, this was "a principled decision to exclude those who challenged the structure of the state from enjoying its benefits."[17] However, the authorities received counsel that such a principle would not be in accordance with European law, and neither, given that many social-security benefits were contributory, would withholding them be legal under British law. However, counsel's opinion on the second method of enforcing compliance, disenfranchisement, was, from the government's point of view, more positive. British law already provided a list of documents, one of which had to be produced in order to cast a vote in Northern Ireland. Legal opinion was that simply replacing this list with a necessity to produce an ID card would be totally lawful.[18]

The authorities foresaw opposition to this scheme from both sides of the Northern-Irish sectarian divide. Nationalists would feel targeted while unionists would feel aggrieved at being treated differently to others in the United Kingdom. This was, of course, another reason why the ID card scheme was introduced through a system of "low-key diffusion" by being coupled to the existing driving licence.[19] However, even the scheme's advocates saw that it would also entail two other sets of problems. The first of these was associated with its administration. Though the RUC wanted ID cards introduced, it did not want to become the issuing authority. This would, the force argued, take up hours of police time and congest police stations, thus making officers' lives even more dangerous. Nevertheless, the scheme's advocates argued that its administration could not, given the security situation, be conducted by anyone other than the RUC.[20] Second, there were the issues of foreign residents in the area and of visitors to the province. This later raised the vexed question of what would happen were visitors from another part of the United Kingdom to go to Northern Ireland? Even the scheme's proponents saw that it would be "presentationally unattractive" to demand that such people should have to carry a passport to visit another part of their own country.[21] But there was also the issue of visitors from the Republic of Ireland. The Irish government showed no sign at all of wanting to introduce its own ID cards, never mind participate in a British scheme, and the security forces were aware that republicans in Northern Ireland could either secure identity documents from the Irish Republic, or claim to be visitors from there, thus throwing a spanner into the works of the whole plan.

Framed in this way, these are issues characteristic of an attempt to introduce ID cards in Northern Ireland, but similar issues around the security of issue and the place of foreign nationals would arise when ID cards were discussed for the rest of Britain. Moreover, as in Northern Ireland, there would also be the vexed issue of the carrots and sticks employed to cajole or compel the public into obtaining and using the cards. Associated with this, there was also the question of whether an ID card should be compulsory. In the end, this discussion about Northern Ireland did not lead to the imposition of ID cards in the province because as Tom King, the secretary of state, explained to Margaret Thatcher, compulsory cards would make a bad political situation worse while a system of voluntary cards would not be up to "doing the job we have in mind."[22] However, the issues of driving licences and ID cards would resurface a little over five years later.

3

Margaret Thatcher left Downing Street in late November 1990 and her successor, John Major, won an election in April 1992. It was just under two years after this victory that the issue of reformatting driving licences resurfaced and, when it did, it would come already intertwined with suggestions of ID cards. Thus, early in 1994, William Waldegrave, the chancellor of the Duchy of Lancaster, wrote to the minister of transport, John MacGregor, acknowledging his plans to "expand the role of the driving licence as an ID document rather than just a road permit," which was to be achieved by the introduction of California-style licences for non-drivers.[23] Waldegrave noted that this development was being forwarded at the same time that Peter Lilley was, as part of his campaign against fraud, moving towards introducing a plastic card for social-security claimants, and he (Waldegrave) speculated whether there "would be scope to combine on the one card the requirements of the DVLA with those of Social Security and perhaps other Departments and Agencies as well."[24] John Major gave a cautious welcome to these proposals being content to allow work to proceed while waiting for an update from Lilley and reserving judgement on the topic of California-style voluntary ID cards. He was however, prepared to repeat his "general readiness to consider identity cards to reduce fraud."[25] In his original suggestion to the government's Home and Social Affairs Committee, MacGregor had proposed setting up a ministerial group to examine the whole topic of smart cards. Lord Wakeham subsequently wrote to Major to tell him that: "there was a strong view in the committee that rather than individual departments proceeding separately ... there would be considerable benefits in taking an overall look ... to make sure that action was coordinated as far as possible."[26] Major agreed with this approach and gave the go ahead for the foundation of a ministerial committee to look at the subject.[27] Wakeham would chair this committee, the Ministerial Group on Card Technology, but before it had even convened, he received a suggestion from David Hunt, the secretary of state for employment (his department was working with Lilley's on "a number of ways of combatting benefit fraud") that: "there might be some scope for a national ID card to help in this area, as well as in other areas of Government."[28] Thus, an examination of card technology was, from its inception, geared towards interrogating its fullest possible use as an ID card and this would create the political space for their advocacy.

This committee had a high-powered membership. Lord Wakeham, its chair, was the lord privy seal, and the committee also contained: the home secretary, Michael Howard; the lord president of the council, Anthony Newton; the chancellor of the Duchy of Lancashire, William Waldegrave and the secretaries of states for: transport, social security, health and the treasury (John MacGregor, Peter Lilley, Virginia Bottomley and Michael Portillo respectively).[29] It first met on 24 May 1994 and was wound up, with its residual functions being absorbed by the Ministerial Committee on

Home and Social Affairs, on 18 July 1995.[30] During its fourteen months, the Ministerial Group on Card Technology aired all the arguments about population data and ID cards that had been unleashed by the proposal to reformat the driving licence. In the process, it also provided the launch pad for a May 1995 green paper, authored by Michael Howard, on the subject of ID cards, which duly led to a House of Commons Home Affairs Committee report in June 1996. Howard's publication was remarkable, given that its author was a strident campaigner for ID cards, in that it was a genuinely consultative document. The exploratory nature of this publication was a product of the politics of the committee where there had been considerable opposition to the introduction of a pan-department smart ID card. The nature of this opposition and the issues raised in the committee's discussions more generally are the subject of this section.

The committee had five options before it and its job was to recommend one of these. The paths it could suggest the government follow were: first to do nothing extra. Such a choice would allow departmental plans currently in train to run their course. Second, to allow existing plans to develop but add, over and above these, a simple "dumb" national identity card. Third, leave existing plans alone, as with the previous option, but introduce a smart national identity card. Fourth, allow the benefits card to be issued and then transform this, at some unspecified date, into a multi-function government smart card. Fifth, follow the same pattern as the fourth option, but by piggy backing on a reformatted driving licence.[31] Thus, from the beginning of its work, the committee discussed the potential use of smart cards by the government. To avoid any confusion, such a card was defined, in June 1994, as being a photo card with a magnetic strip and an embedded processor and possibly, an optical storage area on the card's surface.[32] The committee was presented with a list of twenty-two functions that a smart card could perform in the provision of services to the public, spread across six different areas of interest to government. These included road tolling, paying on public transport, reducing social -security fraud, storing medical records, holding electronic credits for training programmes and acting as a residency permit for immigrants.[33]

Michael Howard was keen to stress to John Major that the technology existed to make a smart card a safe and secure means of holding all this data.[34] However, even his assessment of an ID-card scheme raised considerable problems and, as was touched on when the issue was raised regarding Northern Ireland, the crux of the problem was in verifying the identity of the applicant at the point of issue. A passport was obtained, for the first time, by providing a birth certificate, to prove nationality, along with two photos of the applicant that had been countersigned by a professional person who had known the applicant for at least two years. Measures were taken to check the reliability of these "countersignatories," but the fact that these were needed at all indicated that there was a loophole in the system's security. To close this, Howard advocated the use of biometric data, such as encoded fingerprints, on the new ID card. But even he admitted that this would only make it "more difficult," not impossible, for a person to get round the system and adopt multiple identities. Thus, even with this addition, there would still be a need for "administrative checks on identity."[35] In this, it is possible to discern how, rather than providing some sort of definitive resolution to issues raised by the biopolitical relationship between the government and the people, the pursuit of knowledge of the population actually opened the gates to the chimerical pursuit not only of perfect data, but also of perfect security for the data systems themselves. Moreover, none of this exposition paid anything more than scant attention to the likely reactions of the public to being compulsorily fingerprinted.

When the police gave evidence to the House of Commons Home Affairs Committee, the representative of the Association of Chief Police Officers stated that: "we are anxious to say we are not the drivers of this debate." Rather the police had "found ourselves sucked into the debate by other peoples' propositions."[36] This statement may have been true insofar as the police did not actually propose the introduction of ID cards in 1994. Nevertheless, this should not obscure the force's longer-term advocacy of ID cards, or indeed, their support for both compulsory fingerprinting, and the mandatory carrying of driving licences. Moreover, even if it is accepted that the police had to be dragged into the debate at this point, once they were before the committee, these officers were only too willing to express an opinion. When they did so they, unsurprisingly, concurred with Howard "that the use of a fingerprint in a coded form" would be the best biometric system to provide security on an ID card. However, the committee heard from another witness that: "the fingerprint has this criminal connotation which I think we have to bear in mind when dealing with the public."[37] Moreover, the Ministerial Group on Card Technology was fully aware of this latter point. At its first meeting, the point was forcefully made that: "some members of the public already have considerable reluctance to provide the police with fingerprints where it was in their interests to do so (for example, where their houses had been burgled)."[38]

The Ministerial Group on Card Technology further echoed the debates in Northern Ireland when it moved on to discuss whether any national ID card should be compulsory or voluntary. Michael Howard noted that there were arguments to be heard from both sides of the debate but stated that: "my personal preference, based on an assessment of the likely benefits to my own department, is to move straight to a compulsory scheme."[39] On this point, concerning a compulsory ID card scheme, he would find himself, perhaps surprisingly, aligned against the police. The chair of the Police Superintendents' Association, Brian Mackenzie, agreed with the chair of the House of Commons committee that such a scheme would

> certainly make our job a lot easier and it may well be that answers we are giving are surprising, but … we do not want to live in a police state and we do not want to organise a police state or police a police state.[40]

The home secretary did not spend any time at all in his fourteen-page memorandum dwelling on the civil-liberties case against ID cards, but rather highlighted the costs to government of a compulsory scheme. These included non-financial costs that would encompass a deterioration in police-community relations, which were "likely to be felt most severely amongst ethnic minority communities particularly if the identity card was to be used for immigration purposes."[41] There were also, in financial terms, the costs of monitoring compliance and prosecuting offenders along with the costs involved in establishing and running an ID card scheme. These, Howard pointed out, would be higher were a card to be compulsory, because in such circumstances, the government would find it difficult to make people pay to have one. Because people had always paid for a driving licence, by using this document as a gateway to an ID card (a voluntary system) and so reducing the government's costs, the government could not be accused of taxing people.

By the time the House of Commons Home Affairs Committee considered the issue, Howard had already been obliged to give up lobbying for a compulsory scheme and had published a green paper that simply set out the various arguments. However, the parliamentarians acknowledged that being required by law to have, carry and produce on demand an ID card "is thought by some to be a reduction in the civil liberties of the citizen" and stated that: "we do not recommend the introduction of a compulsorily held card."[42] In agreeing with the introduction of a voluntary scheme, this committee raised another issue seen in Northern Ireland. This was, as the military had always maintained, that an ID card could only function if it were used universally. Thus, one of the main factors in introducing a voluntary card was the need to build into the plan a means for either cajoling or enticing people into getting one. In the case of Northern Ireland, the authorities were willing to discuss stringent means by which to coerce the people into using an ID card. With this national scheme, measures such as disenfranchisement were not debated, but attention did turn to how to make the cards as attractive as possible in order to boost their uptake and use. This, of course, was a form of indirect compulsion and when it gave evidence to Parliament, the Consumers' Association was not alone in noting that:

> it is clear from the experience of other countries ... that a voluntary scheme rapidly becomes effectively compulsory because the card is so frequently demanded by public and private sector bodies in circumstances where other forms of ID had previously been adequate.[43]

This point, about how a voluntary ID card was in fact not voluntary at all, was made even more forthrightly eight years later when, under Blair's government, attempts were made to resurrect Howard's scheme. Here, giving evidence to the House of Commons Select Committee on Home Affairs, the Law Society characterised the word "voluntary" as "very disingenuous since anyone renewing a passport or driving licence would have no choice but to register, and pay, for an ID card version of these documents." Moreover, at the same committee, the information commissioner "similarly talked of an illusion of choice." All of which led the committee to conclude that describing the government's plans as voluntary "stretches the English language to breaking point."[44]

This was, of course, precisely why some advocated the voluntary system. This incremental approach would, for example, allow the card to establish its usefulness in a few areas of life before its facilities were extended to embrace others.[45] The scope of the areas of life where the public could benefit from having one of these cards was thus elastic. As the House of Commons committee noted, these could stretch from driving, using public transport or the local library, to banking, retail credit and "plastic money" along with "others not yet devised."[46] In addition to accepting services provided by a smart ID card, there was also the question of whether the public would accept the card as such. This concerned the argument aired around the late 1980s proposals for U.K. driving licences (above) and whether Britain was a "credit-card society."

When, in September 1994, Michael Howard wrote to John Major stressing the safety of his proposed smart ID cards, he also maintained that "plastic cards are now common place," and that photo-bearing cards were equally increasingly the norm.[47] On this point, the transport secretary, Brian Mawhinney (he had taken over from MacGregor in July 1994), wrote to Major a few days later also lobbying for this proposal, and in doing so, cited figures of 85% public approval for the new licence.[48] Howard used this positive public response as proof that: "the time is right to test the water" regarding the introduction of ID cards.[49] The ID card was always touted as something that would make interactions with the government more convenient for the public, and it seemed as though the public were ready to accept the form through which they might reap this convenience. However, the notion that Britain had turned into a credit card society was not one that was accepted by all those on the Ministerial Group on Card Technology. Thus, at the group's first meeting not only was the British people's antipathy to giving up biometric data emphasised, it was also asserted that this may have been part of a broader social pattern as "the use of credit card services was, for example, by no means universal."[50]

If the British as individuals were not yet as modern as the population of the United States, and so were less likely to understand and accept the conveniences offered to them by an ID card, then those advocating the new ID card could always claim that the card would still appeal to the public's political inclinations. In doing so, the arguments would be made that an ID card would reduce immigration and fraud. Howard argued that a "well designed" ID card could be used to catch those who were in the United Kingdom illegally when they came into contact with any part of the state. Furthermore, it could also be used to prevent employers offering work to those who were in the country illegally.[51] In other words, it would create a hostile administrative environment for immigrants. This, he argued, "could be developed and implemented effectively and with public support."[52] Mawhinney drew attention to what he described as "burgeoning prescription charge exemption fraud," while a June 1994 meeting of the ministerial group heard about Powellite-style issues of "social security fraud and false claims of entitlement to National Health Service treatment."[53] All these issues featured prominently in Howard's green paper where he also added the issue of electoral fraud, which, until this point, had not been mentioned anywhere in these discussions.

For its advocates, the potential uses of a smart ID card were not limited to government services. In fact, they saw the device as possibly supporting the provision of a wide variety of services by private sector organisations, such as banks and shops. One of the reasons for introducing these cards through a driving licence was that people paid for a driving licence and, following this line of reasoning, Howard suggested to John Major, that given how the cards would benefit business, businesses might well be expected to become "partners in developing, funding and using such a card."[54] He held this view despite the fact that it had been argued in government circles that the public's considerable disquiet over the linkage of government datasets "would be still greater if the private sector also had access to data held on the cards."[55] The government commissioned a study by its Office of Public Service and Science into how a smart ID card might work and, in the summer of 1994, this body reported to the ministerial card technology group. This report had been written as a result of research undertaken by a consortium of four private IT/communication systems companies. They proposed that a single commercial organisation be created to run the scheme on behalf of both the government and any business that wanted to use the card. The consortium would fund the establishment of the scheme and would derive its income from, amongst other sources, "an entry charge for departments and commercial organisations wishing to join the scheme." This, it argued, would deliver benefits to government departments in the forms of clearer communication and simplified administration. It would also offer benefits to private businesses, which: "would have the opportunity to offer a range of value added services to existing card holders."[56] In other words, this proposal, from within the government, would have set up a national, compulsory data bank, which would have become the centre of a data market.

However, though business may have wanted to get its hands on government data about the population, it wanted nothing to do with this ID card scheme. The House of Commons committee reported that the British Bankers Association foresaw major difficulties in putting data from more than one source onto a single card and was "adamant that private sector and public sector cards should not be linked." Organisations representing other sectors of banking and commerce concurred with this view, as did the British Retail Consortium.[57] The government had always argued that its data-gathering operations in general, and its ID cards in particular would deliver a net benefit to the public, that, in other words, these were a public good. These representatives of private business now turned this reasoning back on the government to argue that: "it would not be right for them to bear the costs of something which Parliament had decided to introduce as benefitting society as a whole."[58] However, when slightly pressed by the committee, one businessperson did agree that in order for an ID card scheme to be attractive to the private sector, it would need to be compulsory, because what business wanted was data on a mass scale. Though these ID cards were not introduced, this point about the ability of government compulsion to harvest data from the entire population and the attractiveness of this as a proposition to business was one that would only increase in importance. This is examined in the next chapter.

The strategic centralisation, standardisation and linkage of government population data that were at the heart of the plans for an ID card drawn up by the consortium of firms did not only appeal to business. In fact, these were the characteristic features of everything proposed as a result of the data turn in government. These had been the essence of Acheson's vision for the NHS, they described the purpose of Wilson's *People and Numbers* plan, they were why Redfern had wanted to reformat the electoral register; they underpinned every common numbering scheme within and across departments and as such they were the key to understanding the centralisation of data institutions that was being finalised by Bill McLennan at this time (see Chapter 4). In this sense then, this ID-card scheme was all of a piece with these broader biopolitical currents. The Ministerial Group on Card Technology had been established to study the developing technology of card systems and to ensure that government departments coordinated their efforts at introducing these systems specifically in order "to avoid unnecessary duplication of such cards."[59] Smart cards, such as the proposed ID cards would, as the House of Commons committee reported, "in practice involve the establishment of a supporting database."[60] Avoiding what Waldegrave termed the "dangers of waste and duplication inherent in continuing to develop cards in a fragmented way," would therefore lead to government issuing fewer cards, which would therefore mean fewer databases.[61] One smart national ID card would consequently produce one database: a population register in all but name.

However, the fact that this type of data drive had been at the heart of government for over thirty years by the time this scheme emerged, must raise the question of why these proposals to introduce an ID card were rejected by ministers. Cabinet-level opposition to ID cards came from two main sources: The Foreign Office and the Department of Social Security. At the Foreign Office, Douglas Hurd, the foreign secretary, was very concerned about the impact that any ID card would have on the common travel area that the United Kingdom shared with the Republic of Ireland. If ID cards were to be issued in the United Kingdom, this would, as has been discussed, raise the issue of the large number of Irish people resident in the United Kingdom who would presumably need to carry passports, or some other documents in order to identify themselves. This would have been "politically sensitive" for Anglo-Irish relations at the best of times, but this relationship was even more sensitive and important than usual to the British government in 1994. This was because the United Kingdom was embroiled in complex frontier-control negotiations with the European Community (EC) in which the Irish were the United Kingdom's only supporter.[62] In these delicate talks, the United Kingdom was attempting to retain "the present system of immigration checks at the point of entry for all those coming into the United Kingdom (except in respect of the common travel area)."[63] The point was that during these talks, the government had consistently argued that it needed to maintain the status quo of border checks because it did not routinely check identity once a person was inside the United Kingdom, indeed it was unable to do so because the United Kingdom did not have ID cards. Any introduction of ID cards, or even a serious plan to introduce them, would thus pull the rug out from under the government's feet on the highly sensitive issue of immigration control. Under these circumstances, it is small wonder that Hurd, was critical of the ID card scheme or, that when he wrote to Major, this issue was the first problem he highlighted with Howard's plan.[64]

Peter Lilley set out the view of the Department of Social Security in a memorandum to the ministerial group in June 1994. He noted that his department had conducted a study into how its system of payment through order books could be improved. This had concluded that sweeping changes were needed and by the time the Ministerial Group on Card Technology started to meet, these changes in the social-security system were underway. These would involve the use of a customer-held magnetic-strip card, of no intrinsic value, to authorise payments at post offices that would be equipped with terminals to read the cards. This project was estimated to cost 130 million pounds and, under the government's Private Finance Initiative, Lilley was looking to set up a consortium of private backers to develop, implement, manage and finance this plan. Lilley argued that the Post Office saw this as an essential part of its business strategy and that involving private businesses in the risk also gave them a veto over what type of card, with its congruent costs and hazards, ought to be used. His department had planned for a roll out of this system in 1996 and any delay while the scheme was retro-fitted to become a national ID card, or was subsumed into a different card scheme, would cost more money lost in fraud.[65]

By the autumn of 1994, Lilley was becoming increasingly concerned about Howard's plans. Indeed, he wrote to the home secretary to express his "profound reservations" about Howard's intention of announcing the ID-card at the Conservative Party's annual conference. Such a course of action would, Lilley wrote, "expose us to considerable political risk for little immediate gain."[66] The social-security card was definitely going ahead and he was concerned that this would get confused with Howard's proposed ID card "and that the political storm we generate would jeopardise" his department's plans. Moreover, all this political fallout could well be generated in order to advance a scheme that might not even function, because no one involved in these discussions could state categorically that the systems envisaged for the smart ID card would work as planned.[67] The government thus ran the risk of ending up with nothing.

Hurd's opposition to the ID card was based on his position as foreign secretary where he was dealing with the frontiers negotiations. It might be argued that when he had been home secretary, at the time of the North Review, he had refused to undertake a government study of ID cards, and thus that it is conceivable that he might have opposed their introduction, regardless of his position in cabinet, on a point of principle. However, this is not what happened since his opposition was based on the arrival of the political contingency in the form of the EC negotiations during his tenure at the Foreign Office. Moreover, as will be seen in Chapter 9, Hurd was not a devotee of data protection in general.

Lilley's opposition to the ID card was likewise driven by his position in government. As has been seen in Chapters 4 and 6, he had clear views about the importance of using data and was driving his department forward in this regard. Indeed, in arguing against the ID card, he wrote to tell Major that his department was "introducing a range of new and standardised procedures for checking identity at the point of claim, and plans are being developed to make better use of the data already held within the Department."[68] It may be ironic, but it is thus the case that Lilley argued against the ID card not because he was opposed to gathering and deploying population data, but precisely because he was in favour of doing so. He had his own departmental plans in place to use population data to tackle fraud and modernise the social-security system. The financial aspect of these had been worked through and he was confident that the technology for them would work. He was prepared to countenance a possible expansion of the benefits card, once it was up and running, but getting this in place remained his priority and, he insisted, should also be the government's focus. Therefore, cabinet opposition to the ID card did not present an alternative of Britishness. This was never a case of Lilley or Hurd standing as the 1990s equivalent of Lord Goddard or the post-war home secretary, James Chuter Ede. The choice presented by the opponents of Howard's scheme was one between his proposed use of population data, or Lilley's data-driven system.

In August 1996, the government published its reply to the House of Commons Home Affairs Committee's report on identity cards. In this, the government stated that it intended to go ahead and introduce a voluntary ID card combined with the new photo-bearing driving licence and to offer a California-style version to non-drivers. There would also be a separate form of the new driving licence available for British people who did not want the ID card provisions or for foreign residents in the United

Kingdom.[69] However, this hotchpotch of cards was never introduced. This was partly because the Conservative government was removed from office in the general election of May 1997, but, more importantly, because the EU ruled against the use of these hybrid driving licence-ID cards. Here, after a long period of gestation, the EU finalised its rules on the standardisation of driving licences. This policy decision made combining the licence with an identity card illegal. Thus, the traditional distaste for documents that was at the heart of Britishness would remain in place as a result of the actions of European institutions.

When it gave evidence to the House of Commons committee, the political and constitutional reform group Charter 88 got to the nub of the identity card issue when it pointed out that:

> as far as we can deduce any logic in the current moves towards a national identification system it is this: that those at the very heart of government are seeking to find ways of centralising and standardising all the information that is currently held on individuals.[70]

Government attempts to do this, whether or not this involved the use of an identity card, were always undertaken to come to an overall view of the individual. This overarching view would then become the accepted definition and classification of the person under consideration. As was seen in Chapter 4, the research conducted for the 1981 census in Harringay concluded that a large number of people misidentified themselves or their family members, all of which clearly indicates that government officials had a view of who individuals/people were and how they should be categorised. Indeed, it was on this basis that the Home Office was adamant that: "in an identity card scheme it is not only important that an individual should have a *single* identity but also a *correct* identity."[71] It was clearly important to the government that people should know who they really were, but it was even more important that government should know this. Moreover, were there to be any lack of clarity on this point, there would be little doubt about which view, that of the individual or that of the data system, would prevail.

Notes

1 James C. Scott, *Seeing Like a State: How Certain Schemes to Improve the Human Condition Have Failed* (London: Veritas, 2020), 77–78.
2 TNA, RG 22/54, *Personal Numbers and Population Registration – Background Notes*, n.d., ca. Feb. 1973, 4.
3 TNA, HO 310/314, Metropolitan Police, *Road Traffic Law Review – Supplementary Report*, 14 Apr. 1986, 1.
4 Ibid.
5 TNA, HO310/314, Angel to Davidson, 23 Apr. 1986, 1.
6 TNA, HO 310/314, Wooldridge to Davidson, 8 July 1986, 1.
7 TNA, HO 310/314, Angel to Hart and Fittall, 26 Feb. 1987, 1 and 2.
8 Ibid.

9 TNA, HO 310/314, Fittall to Angel, 4 Mar. 1987, 1.
10 *The Road User and the Law: The Government's Proposals for Reform of Road Traffic Law*, Cm. 576, Feb. 1989, 3 and 22.
11 TNA, HO 310/314, Angel to Hart and Fittall, 26 Feb. 1987, 1 and 2.
12 TNA, PREM 19/4735, *Draft Consultation Paper*, Annex A, n.d., ca. 1993/94, 1 and 2.
13 Ibid.
14 TNA, PREM 19/3379, *Introduction of Identity Cards in Northern Ireland*, 18 Oct. 1988, 1.
15 Ibid., 3.
16 Ibid.
17 Ibid.
18 Ibid., 6.
19 Ibid.
20 Ibid., 7.
21 Ibid., 4.
22 TNA, PREM 19/3379, King to Prime Minister, n.d., ca. Oct./Nov. 1988, 7 and 8.
23 TNA, PREM 19/4735, Waldegrave to MacGregor, 14 Jan. 1994, 1.
24 Ibid.
25 TNA, PREM 19/4735, True and Neville-Rolfe to Major, 21 Jan. 1994, 4.
26 TNA, PREM 19/4735, Wakeham to Major, 4 Feb. 1994, 2.
27 TNA, PREM 19/4735, Francis to Bailey, 8 Feb. 1994, 1.
28 TNA, PREM 19/4735, Hunt to Wakeham, 22 Feb. 1994, 1.
29 TNA, CAB 130/1486, Ministerial Group on Card Technology, *Meeting*, 24 May 1994, 1.
30 TNA, CAB 130/1486, Ministerial Group on Card Technology, *Dissolution of the Committee*, 18 July 1995, 1.
31 TNA, CAB 130/1486, Ministerial Group on Card Technology, *Multi-Function Government Smart Card Feasibility Study, Memorandum by the Parliamentary Secretary. Office of Public Service and Science*, Feb. 1995, 2–3.
32 TNA, PREM 19/4735, C.C.T.A., Government Centre for Information Systems, *Security Issues in the Smart Card Life Cycle*, June 1994, 4.
33 TNA, CAB 130/1486, C.C.T.A., *Smartcard Report*, May 1994, 9.
34 TNA, PREM 19/4735, Howard to Major, 21 Sept. 1994, 2.
35 TNA, PREM 19/4735, Home Office, *An Assessment of a National Identity Card Scheme*, Sept. 1994, 2–3.
36 House of Commons, Home Affairs Committee, *Fourth Report, Identity Cards*, 1, 26 June 1996, Minutes of Evidence, 19.
37 Ibid., 103 and 82.
38 TNA, CAB 130/1486, Ministerial Group on Card Technology, *Meeting*, 21 Feb. 1995, 3.
39 TNA, PREM 19/4735, Home Secretary, Ministerial Group on Card Technology, *The Case for National Identity Cards*, 27 June 1994, 2.
40 House of Commons, Home Affairs Committee, *Identity Cards*, 26 June 1996, 23.
41 TNA, PREM 19/4735, Home Secretary, *The Case for National Identity Cards*, 27 June 1994, 8.
42 House of Commons, Home Affairs Committee, *Identity Cards*, 26 June 1996, xxiv.
43 Ibid., xxv.
44 House of Commons Select Committee on Home Affairs, *Fourth Report*, 30 July 2004, 29.
45 TNA, CAB 130/1486, Ministerial Group on Card Technology, *Meeting*, 24 May 1994, 4.

46 House of Commons, Home Affairs Committee, *Identity Cards*, 26 June 1996, xv and xvi.
47 TNA, PREM 19/4735, Howard to Major, 21 Sept 1994, 2.
48 TNA, PREM 19/4735, Mawhinney to Major, 23 Sept. 1994, 2.
49 TNA, PREM 19/4735, Howard to Major, 21 Sept 1994, 2.
50 TNA, CAB 130/1486, Ministerial Group on Card Technology, *Meeting*, 21 Feb. 1995, 3.
51 TNA, PREM 19/4735, Home Office, *Memorandum on a National Identity Scheme*, 6–7.
52 TNA, PREM 19/4735, Howard to Major, 21 Sept. 1994, 2.
53 TNA, PREM 19/4735, Mawhinney to Major, 23 Sept. 1994, 2; and, TNA, CAB 130/1486, and, Ministerial Group on Card Technology, *Meeting*, 30 June 1994, 1.
54 TNA, PREM 19/4735, Howard to Major, 21 Sept 1994, 1.
55 TNA, PREM 19/4735, Reynolds to MacNaughton, 27 Sept. 1994, 3.
56 TNA, PREM 19/4735, Ministerial Group on Card Technology, *Further Progress Report*, n.d., ca. Aug.–Oct. 1994, 2 and 3.
57 House of Commons, Home Affairs Committee, *Identity Cards*, 26 June 1996, xvi–xvii.
58 Ibid., xxxvi.
59 TNA, CAB 130/1486, Ministerial Group on Card Technology, *Interim Progress Report on the Card Study*, 24 May 1994, 1.
60 House of Commons, Home Affairs Committee, *Identity Cards*, 26 June 1996, xxix.
61 TNA, CAB 130/1486, Ministerial Group on Card Technology, *Meeting*, 24 May 1994, 1.
62 TNA, PREM 19/4735, Home Office, *An Assessment of a National Identity Card*, 24.
63 TNA, PREM 19/4735, Home Office, *Memorandum on a National Identity Scheme*, 3.
64 TNA, PREM 19/4735, Hurd to Major, 26 Sept. 1994, 1.
65 TNA CAB 130/1486, Ministerial Group on Card Technology, *Memorandum by the Secretary of State for Social Security, Social Security Plans for the use of Payment Cards*, 22 June 1994.
66 TNA, PREM 19/4735, Lilley to Howard, n.d., ca. Oct. 1994, 1.
67 Ibid.
68 TNA, PREM 19/4735, Lilley, extracts in: *Prime Minister's Briefing Notes*, 20 Oct. 1994, 3.
69 *The Government Reply to the Fourth Report from the Home Affairs Committee Session 1995–96*, Cm. 3362, Aug. 1996, 1.
70 House of Commons, Home Affairs Committee, *Identity Cards*, 26 June 1996, 139.
71 TNA, PREM 19/4735, Home Office, *An Assessment of a National Identity Card*, 3. Emphasis in original.

8 Data Systems 1979–97 (3), the Government Data Network, the Information Society Initiative and *government.direct*

After he was elected, in 1964, Harold Wilson set about inaugurating a revolution in government data systems. Once his attempt to deliver this, through the *People and Numbers* population-registration scheme, was stalled, attention turned to reforming the institutions and systems at hand to produce something capable of delivering an improvement in the flows of population data. Some of these attempts, such as the longitudinal study, the centralisation of the Government Statistical Service (GSS) and the creeping usage of common-numbering systems were successfully introduced while others, such as ID cards and a recast electoral register were not. But in November 1996, the government returned to the concept of an explicit full-blown data revolution and published a green paper outlining its plan. This green paper was called *government.direct* (it was usually written like this, in lower case and italicised lettering). This outlined a technologically driven system that, though it had not existed outside the realms of science fiction when Wilson was in Downing Street, is nevertheless best seen as the flowering of everything that had preceded it. Thus, though this plan centred on IT, even the government's National Information Infrastructure Taskforce stated that it was driven by the government's attitude, not technology. This plan was drawn up by a committee chaired by the first secretary of state and deputy prime minister, Michael Heseltine, and was designed to fundamentally reshape the relationship between the people and the government. It was also centrally based on recasting government holdings of population data.

Though this scheme had its roots deep in the biopolitical thinking that came into government in the mid-1960s, it was also a product of more short-term developments and these are examined in the first and second sections of this chapter. The first of these looks at the Government Data Network (GDN). Once this system was in place, it would allow for the exchange of millions of items of personal information annually. It would thus form the framework for all later government population-data initiatives. In the same way that government often claimed that its ID card proposals were nothing other than a logical development of the credit-card society, so it also claimed that its use of IT for population-data purposes merely represented its flowing with the technological tide. However, the second section

DOI: 10.4324/9781003252504-9

investigates some of the ways in which the government actually formed these currents and not only promoted its own use of IT, but also developed this across British society. This discussion centres on a government scheme called the Information Society Initiative. This is important not only for the impact it had on society, but also because *government.direct* was launched as part of this programme. The third and final section of this chapter focuses on *government.direct*.

1

In July 1986, the government's Central Computer and Telecommunications Agency (CCTA: this body was located within the treasury) announced that the Department of Health and Social Security (DHSS), the Inland Revenue, the Customs and Excise and the Home Office, acting under the aegis of the CCTA, intended to study the viability of building a shared data network. The CCTA issued requests for information to these departments and collated all their responses. By the end of the year, the project was sufficiently advanced for the CCTA to produce a service requirement and invite expressions of interest from five companies or consortia. These were all large businesses as the CCTA envisaged an initial scheme coordinating 85,000 terminals across 3,000 locations.[1]

By late 1986, the Home Office was clearly in favour of computerisation and equally of networking files and systems. Nevertheless, this proposal sent the department into a spin as it was forced to decide whether to join the proposed GDN or carry on with the plans it already had in place to build its own integrated digital network (IDN). The problem officials faced was that they were trying to make a decision despite not having all the information they needed. The treasury had delayed plans for a Home Office IDN and so officials did not have a clear picture of its costs, but neither did they know what the GDN would cost, as bids for the contract had not yet even been requested by the CCTA.[2] There was even a lack of clarity about how the costs of the GDN would be apportioned. In January 1987, the Home Office expressed serious concern that the costs for using the GDN might be doled out in proportion to a department's use of the network. Home Office officials feared that since their department was likely to be a big user of the system, it could be disproportionately impacted, in effect subsidising a technological upgrade of other departments. There was even discussion of whether costs would be apportioned to particular sections of participating ministries and so engender intra-departmental as well as inter-departmental conflicts over budgets.[3]

In the absence of this financial information, the Home Office's officials were forced to make a decision on two other bases. Officials were convinced that either scheme, the GDN or a departmental IDN, could ultimately meet the needs of the department, but the first question facing them was: which could deliver this promise soonest? The need to consider the GDN had compounded the delays at the Treasury and several high-profile sections of the Home Office, such as those dealing with prisons, the police or immigration, had all been waiting quite some time for the upgrade to their systems promised by the department's IDN. Thus officials concluded that, on the one hand, "everything depended on timescales," however, on the other hand, these officials were also capable of looking beyond their department and budgetary turf wars to consider the bigger question of government data provision.[4] Thus, in November 1986, they conceded that the GDN was "in the best interests of the government as a whole."[5] While discussions the following January concluded that it would be far better for the Home Office to participate in the GDN, even if its costs were higher than the departmental IDN because doing so would allow the department to reap more benefits.[6] These officials thus abandoned the department's own IDN and "put all their eggs in the basket" of the GDN.[7]

The Home Office had big plans for its data systems. For example, it wanted to centralise all data on prison inmates and make this readable through a network of 600 terminals at 127 sites. It also planned to introduce a national communications network for the Immigration and Nationality Directorate (IND) based on one thousand six hundred terminals at seventy-five sites. Moreover, in 1987, direct communication between Britain's police forces was only possible through the Police National Computer. This meant that individual forces tended to pursue their own plans with the result that: "there is insufficient standardisation logically to exchange even text traffic between forces."[8] The GDN held out the prospect of ushering in a new era of standardisation and so also of increased data usage and sharing across the new network.

This new network was to be a unified system. Therefore, of necessity it would also be a centralised system. A study of the choice in data systems available to the Home Office saw this as an indisputably positive development that was moving with the tendencies of technology. Thus:

> the trend in data communications, and indeed in all other telecommunications services, is therefore towards centralisation of network design ... It follows that, unlike data processing, where the office has pursued a policy of devolution, telecommunications require a central organisation.[9]

This would provide a "unity of command" that would allow for better use of scarce staff skills. It would also introduce economies of scale and lead to the removal of "much confusion, conflict and duplication in the management of the Home Office."[10] Indeed, the only cloud on the horizon of this bold innovation seemed to be that the GDN had "already stimulated public comment on the possibility of exchanging personal data between government departments."[11]

Some of this public comment had come from MPs. Paddy Ashdown, who would become leader of the Liberal Democrats in 1988, asked parliamentary questions about the security of the GDN and its impact on personal privacy. Though, following meetings with the CCTA, Ashdown was reported to be "relatively happy" and "being reasonable," there was a feeling that "the matter was bound to rumble on."[12] Perhaps the most important of these later rumbles came from Eric Howe, the data registrar. In his 1987 report, he turned his attention to the GDN. Here, on the same page of the report where he raised questions about the national credit register (see Chapter 5), he pointed out that the GDN "could lead to the creation of what are effectively massive and comprehensive databases of information, possibly concerning the whole population." Whether this, potentially very serious shift in British political culture, should be created was, he added, "a matter for Government and Parliament" with the public needing to know how the issues around this project were resolved.[13]

Parliament did discuss the GDN. This discussion lasted eight minutes and occurred last thing at night (starting at 10:43 p.m.) on Thursday 25 February 1988 when the paymaster general, Peter Brooke (he had taken over this office from Ken Clarke in July 1987 and held it for the next two years), answered a question from the Conservative MP Alan Haselhurst. Haselhurst listed all the benefits that would flow from the GDN and then expressed incredulity that he had heard people, even other MPs, expressing doubts about the system, adding that some of them had gone so far as to suggest that what it might really imply "is the transfer of data between departments."[14] Brooke was, of course, only too happy to "put the record straight." In doing so, he indicated that the GDN was needed because the number of computers used across government was increasing dramatically. There had been about 65,000 in 1987, but departmental plans indicated that this was set to rise to 150,000 in 1992 and to 240,000 three years later, with the prospect that by the end of the century, government departments might house 350,000 terminals. All this technology was in use for the one clear aim of improving the quality and efficiency of government business, but this was impossible without improvements in communications. This was where the GDN came into play. The network, Brooke maintained, was nothing more than communications hardware designed to facilitate the "sharing of circuits and switches because this makes economic sense." It did not, he emphasised, "provide open access by departments to each other's data."[15]

However, the GDN had not been inaugurated as a finished product, it was always seen as a network that could expand and this was one of the main "wider benefits" the system's advocates were keen to promote.[16] Indeed, one of the first things the GDN Management Board decided, after they took over the project, was to recommend a full feasibility study of how the system could be converted into one that integrated voice and data services.

Technical experts preferred to begin their introduction of IT by establishing a broad pattern of use and argued that precise needs and processes could be developed and incorporated at a later stage. In this, they reflected and reinforced the thrust of the long-standing desire across Whitehall, to create ever wider and more inclusive databases.[17] These tenets were present from the outset of Margaret Thatcher's time in Downing Street. Thus, in July 1981, she announced the creation of an IT advisory panel, made up of outside business people, to be located in the Cabinet Office. In its preliminary study of its subject this panel noted the pressing need for information to be stored only once. However, it added that should this not be possible, there needed to be "rationalised … and refined interchanges of information," between data holders.[18] By 1988, the Home Office's IT Division could note, in a document portentously entitled *Towards the Future*, that an integrated services digital network "had long been recognised as an eventual development from the GDN."[19] By the time Charter 88 gave evidence to the House of Commons Home Affairs Committee in 1996, they were able to say that the GDN "has now linked the Department of Social Security, the Home Office, Inland Revenue and Customs and Excise" and had led to a system where "millions of items of personal data are exchanged between these departments every year."[20] Clearly then, the centralised hardware of the GDN would be used to create centralised systems and sets of data. These would culminate in the system known as *government.direct* that was outlined by the Conservative government in 1996. This system was never introduced, largely because John Major lost the May 1997 election, but as with other abortive plans to build up population-data resources (through a recast electoral register or ID cards, for example), this plan reveals a lot about the government's aims and objectives. Before this chapter examines *government.direct* in detail, the next section charts the government's growing interest in IT.

2

Chapter 5 showed how government used computer technology as a firewall to conceal and protect its population-data-gathering intentions and operations. It also showed how, in 1987, 40% of British people had never even touched a computer. Some of this sizeable minority might have been government officials since, in 1983, the only material held on computer at 10 Downing Street was a list of the names and addresses of the prime minister's correspondents, while the Conservative Party's central office only got its first computer the year before.[21] Within the Home Office, in 1982, the IND reported that the main problem facing its projects for the coming year was "the competition for automatic data processing resources," which were under considerable pressure.[22] The expansion of computerisation in the IND was described as the "triumph of hope over experience" perhaps because, as one frustrated official noted, senior Home Office staff were "not knowledgeable about computers." With the result that the department in general, and the IND in particular, were "living in the nineteenth century."[23] Ten years earlier, in 1972, there had been 160 computer-based tasks planned or operational in central government, a total that had increased by 25% three years later.[24] Throughout the whole of the public sector, there were 1,870 computer systems in 1971 and 3,255 in 1975, whilst the figures for the country as a whole for the same period showed an increase from 6,075 systems to 13,263.[25]

Clearly this changed. Indeed a National Audit Office report of 1991 opened by noting that government owned computer machinery worth over seven billion pounds and was buying an additional two billion pounds worth of stock annually, adding that "many departments are now heavily dependent on information technology to operate their day-to-day business activities."[26] Indeed the early steps in this process of change, though slow and unevenly spread, could even be seen in the figures from 1970s and 1980s (above), and just as Britain became a credit-card society so the British, and the government, became accustomed to IT. But, it is important to note that, just as the credit-card society was nursed into existence by the government through its stance towards the national credit register, so the government had a crucial role to play in getting Britain working on keyboards and screens rather than typewriters and paper. Shoshana Zuboff has built a convincing case to show how since the dot.com bubble of April 2000 the big technology companies have increasingly turned their businesses toward the harvesting and selling of data about the people who use their systems.[27] But what her work omits is any account of how these companies became so big in the first place and, more broadly, the role of government in creating the eco-system of the political economy that allowed, or encouraged, their growth. In the period covered by this book, the British government was very interested not only in developing its own systems, but in spreading the use of computer technology across society.

In 1995, the EU issued an information pack designed to promote innovation and foster the growth of "the information society in the making." This sought to establish pilot projects to stimulate the involvement of people in the information revolution and to integrate such policies in regional development programmes.[28] The Department of Trade and Industry urged the UK's MEPs to support both this initiative and the EU's guidelines for building trans-European telecommunications networks to "use information as a resource."[29] In Wales, the Welsh Development Agency led a campaign, spearheaded by the government's Welsh secretary, William Hague, to both gain status as a part of the EU's piloting schemes and to encourage business to "take full advantage of multi-media networks in order to succeed."[30]

While these developments were underway, the House of Lords issued a report: *The Information Society: Agenda for Action*. This was the first parliamentary select committee report to be published electronically and in the discussions it engendered, it was pointed out that 22% of British people owned a PC, more than double the number across the rest of Europe, but less than in the United States, where ownership was at 33%. Baroness Dean described these developments as "a revolution which has changed and will continue to change society," while Baroness Seear called it "a revolution at least of the same order of importance as the Industrial Revolution." According to the Earl of Northesk "most people, and particularly legislators are fundamentally clueless about what is going on" with regard to these developments.[31] However, while agreeing with the former comments, the government had to disagree on this latter point. In its reply to *Agenda for Action*, the government pointed out that, far from being clueless, it had already taken big steps in the direction indicated by the report. The days when the minister in charge at the DHSS, a huge user and producer of population data, needed officials to painstakingly explain what a data network actually was, as had occurred with Norman Fowler in 1984 (see Chapter 6) had long passed. In fact, the government had launched the cross-departmental Information Society Initiative (ISI). This forwarded a wide variety of programmes designed explicitly to increase the awareness of IT among business, the public and educators. Taken as a whole, the ISI was, the government trumpeted: "the most ambitious programme of its kind ever attempted in a major nation."[32]

The president of the board of trade, Ian Lang, announced the launch of the ISI on 13 February 1996. This initiative was run from the Department of Trade and Industry (DTI) and styled itself as an "Umbrella Brand" encompassing four parallel government programmes. These were the DTI's own business-related schemes (these are not examined here); the Superhighways for Education Initiative, run from the Department for Education and Employment (DfEE); a cross-departmental programme called *IT for All*; and *government.direct* (examined in the next section of this chapter).[33] In 1996, John Major established the Ministerial Group on Information Technology (IT) to oversee the work of the ISI and all its sub-brands, to set the strategic direction of policy and to ensure that IT initiatives were exploited fully in the national interest.[34] Its first chair was the first secretary of state and deputy prime minister, Michael Heseltine, who was succeeded on this committee by the lord privy seal, Viscount Cranborne, four months later in July.[35]

The first of these DTI-ISI sub-brands examined here was concerned with education. There were two reasons why this area of government policy was of central importance to the government's drive to get Britain more acclimatised to the new technology. The first of these was because it would give the next generation the appropriate skillset. Second, because this was a state sector, the scale of state spending involved could potentially have a big impact on British technology firms. The secretary of state for education and employment, Gillian Shephard, made clear to the Ministerial Group on IT that she would have welcomed the involvement of private money in the provision of IT, but also recognised that the government would have to contribute. This may well have been because the scale of the need was clearly pressing. Thus, even though her department's Superhighways Initiative was still gathering evidence (it was due to report in early 1997), she could still point to the reports of the school inspectorate that highlighted the "shortfalls in modern equipment and limitations in teacher expertise and confidence."[36] Thus, while the average number of secondary school pupils per microcomputer had fallen from thirteen, in 1991–92, to ten, in 1993–94, nearly half of the machines in British schools were, at over five years old, obsolete and incapable of running the latest educational software.[37] Moreover, one of the tightest bottlenecks in the system was the supply of IT-literate teachers and these could only be produced as a result of government investment. Thus, as a senior official from the DfEE told the ministerial group, boosting the computer literacy of schoolchildren would need more than machines, it would need political impetus if the plans were to deliver, and it also needed money.[38] This was a government that did not want to increase its spending and, which, by the middle of 1996, was forced to do so as a result of the escalating disaster brought about by the onset of bovine spongiform encephalopathy (BSE, a deadly disease in cattle) in Britain. But even so, and despite being unhappy about the total bill for computers and the associated training for British schools, Michael Heseltine still agreed to take the proposals to the prime minister and to support them through attempts to mobilise private finance behind them.[39]

The next of these DTI-ISI sub-brands was known as *IT for All*. This was a cross-departmental scheme that was launched in December 1996. Broadly speaking, this was an information campaign. The government saw its role as being that of a catalyst, acting to spur the reaction between two other groups: the citizen/consumer and the technology providers. The problem, as the government saw it, was that people bought technology in order to make their lives better, or easier, but they lacked clear evidence of how the gadgetry might be able to do this. They were, as the Earl of Northesk put it, "clueless." Given this, the government set itself the task of explaining, in the broadest terms, how the technology might work to the advantage of the average person at home, at work, at school or as a citizen interacting with the government. It was not the government's role to explain how bits of kit worked, and still less was it to recommend which equipment a person might buy. It was thus not the government's job to pre-empt the market forces at work in the IT sector, rather the scheme should play "an important role in getting the message of the benefits [of IT] across to citizens."[40] *IT for All* would thus be a campaign that sought to "change attitudes and perceptions" and, in doing this, the DTI urged two broad courses of action. The first of these would seek to work on the public through the provision of information and trained intermediaries, but the second would seek to lead by example, by reforming the way government itself interacted with the public.[41]

This was all in evidence at *IT for All*'s launch event. Here, there were displays and demonstrations, the prime minister published a welcome speech online while other speeches were pre-recorded. Additionally, a free-phone information line was opened in January 1997; there was a teletext page devoted to the campaign and a website. The main thrust though was more traditional and featured several million copies of an easy-to-understand essay titled *IT Could Change the Way You Live, Think and Play* along with a twenty-page booklet, *The Guide to How IT Can Help You*, to be sent free to anyone calling the information phone line.[42]

3

The third and final of these DTI-ISI sub-brands was a government green paper published on 6 November 1996 under the title *government.direct*. Rejected suggestions for the paper's name had all been more descriptive (and capitalised) and included: "Self-Service Government," "Direct Access Government," "Direct Service Delivery," "Accessing Government Services" and "Electronic Direct Delivery of Government Services." The title eventually chosen relied for its informative power on its subtitle of: *A Prospectus for the Electronic Delivery of Government Services*.[43] The paper was drafted by the Central IT Unit (CITU) and presented to the Ministerial Group on Information Technology by the chancellor of the Duchy of Lancaster, Roger Freeman (he held this post from July 1995 until the Conservatives lost the May 1997 election). For these purposes, the term "government services" was used, Freeman told the group, to mean "the provision of information, regulation, revenue collection, collection of statistics, payment of grants and benefits and government purchasing."[44]

Government.direct was an overarching scheme to use IT to cut the cost of administration, but it also recognised that savings would only be realised "if the introduction of new technology was accompanied by radical restructuring of service delivery."[45] An unpublished annex drawn up in its drafting stages noted that it was essential to seize this opportunity to radically overhaul the whole of government business. However, though this was to be a revolution pushed through from within government, it was to be one designed around the perceived needs of the consumers of government services. These companies, groups and individuals were, in the government's eyes, becoming increasingly intolerant of government's top-heavy administrative systems that were seen as being run for the benefit of their officials, not for the users/consumers of the systems who, to add insult to injury, were also the ones paying for this inefficiency. Customers increasingly compared these tax-funded services to the leaner, more adaptive systems in the private sector and government systems came off very badly in this comparison. They were, it was maintained, "characterised by form-filling, duplication and the need to deal with many different offices." This situation was, the paper stated, "unsatisfactory and inconvenient."[46] All of this would, as the Home Office's response to the green paper noted, need "a radical change in … the culture of government."[47]

This cultural revolution was designed to confront the torpidity of officialdom with the reality of what the consumers of government services expected, and it was to be presented to the public in similar terms, as offering them benefits as consumers of government services. Thus, in his foreword to the green paper, Freeman wrote that the proposals in *government.direct* would make services "more accessible, more convenient, easier to use, quicker in response and less costly."[48] Earlier drafts of the paper had even suggested that since the use of IT would reduce government's costs, people who used the new systems to, for example, obtain a driving licence, should reap the appropriate consumerist reward by paying less than those who applied for the same document on paper.[49] This was omitted from the published version, but by choosing to focus the green paper on these benefits, ministers also chose not to emphasise other elements of the discussions that had produced *government.direct*. Here, the most noticeable issue omitted from the published version was any discussion of the role the new systems could play in combatting fraud.

Early drafts of the paper and discussions between ministers had made much of how the proposed radical changes could combat fraud. This would be essential, it was argued, because the new systems would engender new opportunities for fraudsters and so, to promote the system to the law-abiding public, it was necessary to reassure them.[50] This was clearly important, but the government was also interested in using the new system to combat existing frauds and Freeman sought opinions from colleagues on the extent to which the final version of the green paper should highlight this issue.[51] An early draft of the green paper contained a relatively large reference to this topic and mentioned how the Department for Social Security used its systems to prevent fraudulent claims.[52] However, some responses to this version did not endorse this approach and maintained that the focus of the green paper should be on the benefits the changes would deliver to the customers.[53] The "prudent" approach thus recommended was the one adopted in the published version of *government.direct*. This critique of the earlier draft may be seen as surprising when it is realised that it came from the Northern Ireland departments that, as was seen in Chapter 7, were not laggard in either combatting fraud or insisting on the use of identification papers. However, here it is important to note that what the Northern Irish offices were criticising was the way such discussions of fraud exposed the "data matching" (linkage) that was at the heart of the *government.direct* proposals. Officials wanted to keep these linkage mechanisms concealed behind all the presentations of the planned ease of access, for the consumers of government services, that would flow from *government.direct*.

The green paper presented outlines of how the government's systems communicated with the public in the form of a simple diagram. This showed information flowing up to officials in parallel lines that did not communicate with each other at any point. This was known as a "stovepipe" system.[54] What *government.direct* sought to do was to build connections across these flows. In fact what it wanted was to unite them so that people would access government, rather than a particular government service. In this sense then *government.direct* was seen as one-stop shop through which all departments would have access to all the information they needed, whenever they needed it, without needing to ask the public to duplicate data that had already been provided. On this basis, data sharing was, as Freeman wrote, "central to much of the rationalisation the Green Paper proposes."[55] This might require careful handling, but in the ministerial group, there was a consensus that the public could be swung behind this scheme if it were made clear to them how it would be of practical help in their dealings with the government.[56]

The government responded by bringing forward an argument that was similar in many ways to the concept of the task that had been used (see Chapter 2) to undermine confidentiality guidelines within the health service in the early 1970s. Thus, what the government did, in its response to comments it received on the green paper, was to suggest that linking data could assist a member of the public by providing "episode-based" transactions. Here the example given of an episode was that of the death of a relative. It was shown how a system focused on "episode-based" transactions would allow a bereaved person to access all the services they needed at once, removing some of the burden of stress, because the data needed by the providers of different elements of the required service package would be shared between them.[57] This vision was very much in tune with that of the British Computer Society, which, in its response to the green paper, argued that: "we need systems that get back to treating real world objects (e.g.: people) as the single real world entities that they actually are."[58] This in turn carried clear echoes of the whole-person approach forwarded by the framers of the Joint Approach to Social Policy in the 1970s. In other words, the ideas in *government.direct* were a direct continuation of the thinking that had underpinned all biopolitical population-data gathering since the mid-1960s.

The government's public statements on the systems underpinning *government.direct* followed the well-worn path trod by many government speakers in the earlier years, such as Peter Brooke in the February 1988 token parliamentary discussion of the GDN, and so stressed that this new development would not usher in a dystopia of panoptic surveillance. In a parliamentary statement on the project Freeman was "happy to assure the house that fears … that our aim is to … create huge databases to spy on the citizen" were "utterly groundless." The green paper itself argued that: "the government has no intention of merging all the personal or sensitive information held on an individual … into a single database."[59] Interestingly the Home Office raised a cautionary note here, and seemed to call these assertions into question by suggesting that, once the links envisaged were established, the public would increasingly refer to the new way of doing things by the decidedly Orwellian sobriquet of "the system."[60]

However, the Home Office's response to the green paper also raised a much more central and important point. This was that the anticipated savings in administrative costs would only materialise if data were held centrally within a linked system that all departments could access and which would enable them to extract and add information whenever they wanted. This ability would, in turn, necessitate strong central control over the system not least to ensure that all data held by departments was in a standardised format so that the system could process it and other officials could access and use it.[61] Indeed in the discussions and papers generated through the processes that led to the green paper, Freeman made references to "the new inter-connectivity between departments" and how the new system would encourage "greater inter-operability between departments."[62] He also advocated "Mining Government Information," by which he meant drawing together stocks of data that the government held for administrative reasons to make it available through systems designed to link it. He also suggested that there "may be scope for organising information into one or more government-wide database(s)" to meet the needs of business. Moreover, in all this, he concurred with the Home Office's assessment that such developments would need far greater standardisation of data than had ever been the case. Indeed, he went so far as to suggest that even if datasets were not shared at the moment, it would be a good idea to standardise their formatting just in case they might need to be linked in the future.[63]

Though these sets of government population data might be linked in either the short or the medium term, it was a fact that Freeman and the government were keen to broadcast, that they would remain institutionally separate. It was this that allowed him to dismiss notions that he was constructing the infrastructure of an Orwellian nightmare and to repeatedly emphasise that: "existing government databases will generally remain separate."[64] However, this was somewhat disingenuous, because the degree of linkage envisaged between datasets, lubricated by the standardisation of their formatting, would mean that, in parallel to their individual identity, the system would create a new meta-corpus of information that was greater than the sum of its parts. The green paper made much of how people would access the system. This would be done through using an "electronic signature" that could be as simple as answering a selection of previously entered security questions. All of which, the paper reassuringly told the public, was exactly the method they were accustomed to when they used telephone banking.[65] However, what the green paper completely omitted was any discussion of how the system would access its constituent elements. Tim Holt, the first head of the Office for National Statistics (he had previously been the chief statistical officer and head of the GSS having succeeded Bill McLennan in these posts, he was head of the ONS from its formation in 1996 until 2000) pointed out that this systemic linkage would need what he called "a central translator," which could convert between all the different reference systems in use along Whitehall.[66] In other words, achieving the data integration that underpinned these changes would need what was, in all but name, a centrally held common-numbering system: it would need what had always been seen as the spine of any population-registration system and the death knell of the "British citizen's traditional anonymity."[67]

However, the data that might be linked through *government.direct* was not confined to that held by the government. In Chapter 7, it was shown how Michael Howard, and other advocates of ID cards, wanted to introduce a smart card that could be used not only for identification, but also to access services provided by private sector organisations such as banks. The banks wanted nothing to do with his scheme, but that did not mean that private business was uninterested in marketing data. As was seen in Chapter 5, as early as the first half of the 1990s, the government acted to facilitate a creeping commercialisation of NHS data. Moreover, the idea of merging state and privately owned datasets reappeared here in the thinking behind *government.direct*. The system was always presented as one that would make life easier for the citizen. But sometimes people might become involved in episodes that involved both government and private sector organisations. The case in point was a car owner applying for a road tax disc (a paper item displayed in all road vehicles), a process that necessitated the production of valid insurance and MOT certificates. This transaction was usually completed in person at a post office but, were it to be available through the system, the government would need access to the data held by the MOT provider and the insurance company.[68] This prosaic example was, however, clearly the tip of the proposed data iceberg as the government discussed linking its systems to the data of banks in order to facilitate its campaign against fraud in the social-security system.

At the same time as the government was working on this set of proposals, it was also trying to steer a Benefit Fraud Bill through Parliament. This would have allowed the Department of Social Security to increase both the depth and breadth of its data linkage, and it was partly to avoid muddying the waters for this bill that Freeman argued the green paper should not overly emphasise its advocacy of data linkage. But even so, this idea of linking government and private sector databases, was related to the social-security system's increasing clampdown on fraud. Thus, documents circulated in the build up to the launch of *government.direct* made clear reference to government systems being able to access data held by banks and insurance companies to prevent fraudulent claims being made for public money. As had occurred in the case of ID cards, private sector organisations did not trust the concept of such a multifunctioning system and so refused to participate in it. However, at the drafting stage, no mention was made within government circles of what these organisations might expect, or be offered, in return for providing this service to government.

The green paper suggested that people might access government services through the system by means of security questions, but its preferred method of admitting people was for them to have, and use, a smart card. Early drafts of *government.direct* stated that a smart card used for accessing services could also function as an ID card and/or a driving licence with the added possibility of its being turned into a multi-function government smart card at some future date. All such discussions were removed before the paper was published as Howard's plans for an ID card were watered down into the voluntary scheme proposed by the government (see Chapter 7). An early draft of an annex to the green paper also suggested that each card would bear a unique identification number, inserted at the point of manufacture, to which would be added unique numerical identifiers of its registered legitimate user. The system would, in other words, number each cardholder and cards would additionally carry biometric information on the holder for verification processes. The card was not described here as carrying the holder's name and photo but from the point of view of a computer-operated system, this would be an ID card.[69] Indeed in responding to the, much watered-down presentation of this issue in the green paper, the National Association of Data Protection Officers pointed out that: "there is a risk that smart cards may become identity cards." The British Computer Society (BCS) went further than this to state: "it is likely that such a card would be, in effect, an identity card."[70]

It was shown in Chapter 5 how the public may have responded to government approaches for serious political dialogue and subsequently it has been seen how this dialogue was not forthcoming from the government when it came to its proposals for ID cards. This was also the state of affairs during the development of *government.direct.* In fact, the only mentions of consulting the public during these deliberations came from technical experts. Thus a CITU paper, presented to the ministerial group by Freeman, noted that the proposals to establish links between government data systems opened the way to widespread data matching and even though this could be effective at detecting fraud, it was "most controversial" and merited "wider debate."[71] The BCS went further on this point to argue that in addition to being engaged in debate, the people should be empowered and given a veto over which non-government organisations government data was shared with.[72] There is no record of any government minister raising these issues in discussions. It might be argued that instigating such a debate was the whole purpose of the green paper; after all, Freeman wrote stating exactly this on the title page of the document. However, the government was aware that the launch of its *IT for All* campaign had failed to attract media attention and relying on ordinary people hunting out copies of, and dutifully reading, a green paper was, if it was nothing else, somewhat optimistic. After all, nothing had noticeably changed since it had been drily noted by an internal Whitehall enquiry in 1971 that: "the reading of parliamentary Acts is not yet a common pastime."[73]

However, the public was not offered a dialogue about this cultural revolution in British politics, and neither were people offered a veto over the way this might impact what was called, by the associations representing all tiers of local government, the "right to remain anonymous," or what Charter 88 labelled "our right to be left alone and to enjoy ... individual dignity and autonomy."[74] Rather, what was on offer from the government was a series of technical reassurances that the system would not, and indeed could not, be misused. This section of the government's response to the comments it received on the green paper began by rather peevishly stating that: "despite the assurances given by the Green Paper" some of the responses received expressed scepticism about the government's intentions. Further, some even seemed to suggest that the system would become a window into the souls of the people through which anyone in Whitehall could stare. Here, the government followed a twin-track defence of its proposals. First, it deployed what had, by this point, become its long-established, standard defence against all these charges, and asserted that technical systems would be in place to prevent any undue access to, use of, or tampering with the data stored on its computer systems. Second, the government retorted that such prying "will not happen. The law does not allow it and the government will make sure that the law is enforced."[75]

The safety of government computer systems, and thus the reliability of these technical assurances and the attitude of the public to the computerisation of government data processes are examined in the next chapter. But in this reply, the government also placed great store on the law, and seemed to imply that this provided an unchanging, unassailable bastion to protect the rights of Britishness. There are two things worthy of note here. First, this use of the law as a defence against these charges from respondents did not include buttressing the Data Protection Act, which was explicitly judged to be unnecessary in this context, whereas in reality, as was seen in Chapter 5, the British system of data protection was not really fit for purpose. Second, far from strengthening the law around data, and far from regarding the laws that did exist as an unchangeable fact of political life, Freeman clearly understood that *government.direct* could not function unless the law were changed. What he had in mind here were provisions in extant legislation that ring-fenced datasets with confidentiality protocols. In order for data to be shared/linked through the proposed system, any such legislative barriers would have to be removed. However, Freeman did not think that doing this would be problematic. In fact, in his opinion, it would, "amount to no more than removing or modifying some sectoral statutory barriers related to the purpose of particular databases."[76]

What Freeman meant by this dismissive reference to modifying "some sectoral barriers" was, in reality, the dismantling of the remaining confidentiality barriers around Whitehall datasets that would, if left in place, prevent the total linkage of government population data. That these removals were necessary indicates clearly that such linkage was what *government.direct* was really about. As such, this entire proposal is also indicative of the way in which population-data-driven biopolitical thinking had embedded itself in the mind of the British government system. Freeman and the others in the Ministerial Group on IT were all Conservatives, and yet, they all endorsed a system that would have totally changed the nature of Britishness as it applied to the government's use of data about the people. Which is to say that this system would indeed have brought about a cultural revolution.

The government was pleased with the responses it received to the green paper and planned its next moves. These included public "hands-on" piloting events where people could try out the new features, along with the provision of the technology to rural post offices where people would be able to use it. Discussions with groups that had made substantial comments on the green paper were set to continue and the government aimed to have a white paper drawn up by the end of 1998. The government felt that if the take up of the new system followed that of the public's adoption of ATM/cash machines, then the system would be handling 25% of its transactions with the people by 2003.[77]

However, the plan was never realised. Wilson's data revolution had been stymied by a popular privacy campaign. Michael Howard's push for ID cards had been shunted into the sidings by political opponents within the Conservative government and ultimately derailed by the EU, and this re-incarnation of the idea of a drastic, modernising overhaul of British political culture was itself thwarted: by a general election in May 1997 that replaced John Major's Conservatives with Tony Blair's Labour Party. This transfer of power may have changed the personnel in power and some of the policies they pursued, but the desire for population data did not falter. The transfers of power between Wilson and Heath, the moves from Wilson to Callaghan, Callaghan to Thatcher, or from her to Major had not altered this trajectory at all. In fact, as this book has shown, the political imperative behind data gathering actually increased in strength over this period, and this last shift of political power would not lessen it either.

Notes

1 TNA, T 447/1881, CCTA, *Government Data Network (GDN): The Next Step*, 4 Dec. 1986, 1.
2 TNA, HO 524/3, Home Office GDN Advisory Committee, *Meeting*, 3 Nov. 1986, 2–3.
3 TNA, HO 524/3, Home Office GDN Advisory Committee, *Meeting*, 19 Jan. 1987, 2.
4 TNA, HO 524/3, Home Office GDN Advisory Committee, 3 Nov. 1986, 2.
5 Ibid., 3.
6 TNA, HO 524/3, Home Office GDN Advisory Committee, 19 Jan. 1987, 2.
7 TNA, HO 524/3, Home Office GDN Advisory Committee, 3 Nov. 1986, 5.
8 TNA, HO 524/3, *Study of Data Network Options for the Home Office*, 5 Jan. 1987, 4–5.
9 Ibid., 22.
10 Ibid., 23 and 25.
11 Ibid., 13.
12 TNA, HO 524/3, Home Office GDN Advisory Committee, 19 Jan. 1987, 1.
13 The Data Protection Registrar, *Third Report of the Data Protection Registrar* (London: HMSO, 1987), 5.
14 Hansard, *Parliamentary Debates, House of Commons*, 128 (1988), col. 551.
15 Ibid., cols 552 and 554.
16 TNA, T 640/652, *The Government Data Network*, n.d., ca. Apr./May 1988, 1.
17 TNA, HO 524/1, IND, *Computer Strategy Meeting*, 21 July 1982, 2.
18 TNA, CAB 164/1696, *A Preliminary Study by the IT Advisory Panel*, n.d., ca. Dec. 1982, 2 and 7.
19 TNA, HO 524/3, Buck to Moriarty, *Towards the Future – ISDN*, 27 Sept. 1988, 1.
20 House of Commons, Home Affairs Committee, *Fourth Report, Identity Cards*, 1, 26 June 1996, Minutes of Evidence, 140.
21 TNA, PREM 19/2219, Butler to Patten, 10 Feb. 1983, 1; and TNA, HO 411/50, Chesterton to Davidson, 11 Nov. 1982, 1.
22 TNA, HO 524/1, IND, *Computer Strategy Meeting*, 14 July 1982, 1.
23 TNA, HO 524/1, *ADP Presentation*, 5 Nov. 1982, 4 and 6.
24 TNA, HO 411/18, *Government Computers and Privacy, Draft Report to Ministers (2), Note by the Secretary, iii Survey of Personal Information Held in Government Computers*, n.d., ca. Jan. 1972, 1; *Computers: Safeguards for Privacy*, Cmnd. 6354, 16 Dec. 1975, 4.
25 *Computers: Safeguards for Privacy*, 38; *Report of the Committee on Privacy*, Cmnd. 5102, July 1972, 178; *Computers: Safeguards for Privacy*, 34.

26 National Audit Office, *The Management of Information Technology Security in Government Departments*, 26 Feb. 1991, 1.
27 Shoshana Zuboff, *The Age of Surveillance Capitalism: The Fight for a Human Future at the New Frontier of Power* (London: Profile, 2019).
28 TNA, BD 41/445, European Commission, *Information Pack, Information Society*, 1995, 4.
29 TNA, BD 41/455, Department of Trade and Industry, *European Parliament Updating Briefing for UK Members*, 5 Dec. 1995, 2.
30 TNA, BD 41/455, Cutting from: *The Western Mail*, 14 Nov. 1995.
31 Hansard, Parliamentary Debates, House of Lords, 564 (1995), col. xxx.
32 *Information Society: Agenda for Action in the UK. Government Response to the Report by the House of Lords Select Committee on Science and Technology*, Cm. 3450, Nov. 1996, 3.
33 TNA, CAB 130/1517, Official Group on IT, *ISI: Development and Direction*, 9 Jan. 1997, 2.
34 *government.direct: Electronic Delivery of Government Services*, Cm. 3438, Nov. 1996, Foreword, 12.
35 TNA, CAB, 130/1510, Ministerial Group on IT, *Composition and Terms of Reference*, 4 Mar. 1996, 1; and Ministerial Group on IT, *Meeting*, 15 July 1996, 1.
36 TNA, CAB 130/1510, Ministerial Group on IT, Gillian Shephard, *Multimedia in Schools*, n.d., ca. Mar. 1996, 2.
37 Ibid., 2 and 3.
38 TNA, CAB 130/1510, Ministerial Group on IT, *Meeting*, 1 Apr. 1996, 1.
39 TNA, CAB 130/1510, Ministerial Group on IT, *Meeting*, 16 May 1996, 1 and 2.
40 TNA, CAB 130/1510, Ministerial Group on IT, *'IT for All', Supplementary Note*, 28 Mar. 1996, 2–3.
41 Ibid., 3.
42 TNA, CAB 130/1517, Ministerial Group on IT, *IT for All: Progress Report*, 9 Jan. 1997, 1–3.
43 TNA, CAB 202/22, Official Group on IT, *The Direct Delivery of Government Services Using IT*, 27 Sept. 1996, 7.
44 TNA, CAB 130/1510, Ministerial Group on IT, *The Current and Future Uses of IT in Government, Memorandum by the Chancellor of the Duchy of Lancaster*, 17 Mar. 1996, 1.
45 TNA, CAB 130/1510, Ministerial Group on IT, *Meeting*, 15 Oct. 1996, 3.
46 TNA, CAB 202/43, *The Government IT Strategy*, Annex E, 28 June 1996, 2.
47 TNA, CAB 202/314, The Home Office, *Response to government.direct*, n.d., ca. Mar.1997, Annex 2, 3.
48 *government.direct: Electronic Delivery of Government Services*, Cm.3438, Nov. 1996, Foreword, 1.
49 TNA, CAB 202/9, *government.direct, draft 3.0*, n.d., ca. Aug./Sept. 1996, 36.
50 TNA, CAB 130/1510, Ministerial Group on IT, *The Current and Future Uses of IT, Memorandum by the Chancellor of the Duchy of Lancaster*, 2.
51 TNA, CAB 202/22, Official Group on IT: *The Direct Delivery of Government Services*, 2.
52 TNA, CAB 202/9, *government.direct, draft 2.0*, n.d., ca. Aug./Sept. 1996, 35.
53 TNA, CAB 202/38, *Draft Green Paper on the Electronic Delivery of Government Services*, 2 Sept. 1996, 1.
54 TNA, CAB 202/144, Van Orten to Bishop, 6 Feb. 1997, 3.
55 TNA, CAB 130/1510, Ministerial Group on IT, *The Direct Delivery of Government Services Using IT*, 10 Oct. 1996, 6.
56 TNA, CAB 130/1510, Ministerial Group on IT, 15 Oct. 1996, 2.

57 TNA, CAB 202/91, The Government's Response to Comments *government.direct*, 3 Mar. 1997, 11.
58 TNA, CAB 202/244, Wilkes to Simpson, 12 Feb. 1997, 3.
59 TNA, CAB 130/1516, Ministerial Group on IT, *Draft Statement: government.direct*, 25 Feb. 1997, 2; and *government.direct: Electronic Delivery of Government Services*, Cm. 3438, Nov. 1996, 16.
60 TNA, CAB 202/314, Home Office to Bishop, n.d., ca., Mar. 1997, 6.
61 Ibid., 7.
62 TNA, CAB 130/1510, Cabinet Committee on IT, *Memorandum by the Chancellor of the Duchy of Lancaster*, 10 July 1996, 4; TNA, CAB 130/1510, Ministerial Group on IT, 15 July 1996, 1.
63 TNA, CAB 130/1510, Ministerial Group on IT, *The Current and Future Uses of IT in Government, Memorandum by the Chancellor of the Duchy of Lancaster*, 17 Mar. 1996, 2 and 3.
64 TNA, CAB 202/94, *Question and Answer Brief: government.direct*, n.d., ca. Mar. 1997, 4.
65 *government.direct: Electronic Delivery of Government Services*, Cm.3438, Nov. 1996, 24.
66 TNA, CAB 202/100, Holt to Bishop, 28 Jan. 1997, 2.
67 *Report of the Committee on Data Protection*, Cmnd. 7341, Dec. 1978, 264.
68 TNA, CAB 202/43, *The Government IT Strategy*, Annex E, 28 June 1996, 7–8; and *government.direct*, Cm.3438, 29.
69 TNA, CAB 202/56, *Annex to "Government Direct", Data Security*, 12 Sept. 1996, 4–5.
70 TNA, CAB 202/140, Rawlins to Bishop, 5 Feb. 1997, 3; and TNA, CAB 202/244, Allen to Bishop, 5 Feb. 1997, Annex, 5.
71 TNA, CAB 130/1510, Cabinet Committee on IT, *Memorandum by the Chancellor of the Duchy of Lancaster*, 10 July 1996, 6.
72 TNA, CAB 202/244, Allen to Bishop, 5 Feb. 1997, Annex, 6.
73 TNA, HO 411/17, *Government Computers and Privacy, A Report by CSD on Some Manual Systems of Personal Information*, Sept. 1971, 1.1–1.7.
74 House of Commons, Home Affairs Committee, *Identity Cards*, 26 June 1996, 117 and 134.
75 TNA, CAB 130/1516, *Draft Paper: The Government's Response to Comments on the Green Paper government.direct*, 25 Feb. 1997, 11.
76 TNA, CAB 130/1510, Ministerial Group on IT, *The Direct Delivery of Government Services Using IT*, Annex A, 10 Oct. 1996, 3.
77 TNA, CAB 130/1516, Ministerial Group on IT, *Draft Statement: government.direct*, 25 Feb. 1997, 2–3.

9 Government Data Security, the British People and Computers and the 1998 Data Protection Act

The data turn in government spanned the party-political spectrum because it was propelled by the modern political desire for population data to drive biopolitical interventions to underpin policy. However, as the years covered by this book passed, it became increasingly possible to present this as a technological phenomenon. Indeed, this was how government had long sought to portray its data gathering and by doing so it had been able to depoliticise the increasing scope of its data sets. As the previous chapter showed, by 1996, the government was keen to use computer technology to revolutionise all its systems and its relationships with the British people. As part of these processes, it both sought to present systems such as *government.direct* as helpful to the public and, as was seen in Chapter 6, as a safe means of holding and using data. The first section of this chapter interrogates this latter claim of safety and demonstrates how this did not reflect the reality exposed by the government's reviews of its data security. The second section focuses on the British people. This develops Chapter 5's overview of public attitudes to focus on how the British viewed the collection of their data in general and the use of computers to aid this process in particular. This chapter also builds on Chapter 5 by examining, in the third section, the government's reaction to a European data protection initiative that led to the passing of the 1998 Data Protection Act.

1

On 20 March 1989, the Conservative MP Emma Nicholson went to see the Prime Minister John Major, to present him with a report she had compiled. This argued the case for legislation on computer hacking. Nicholson argued that there were 20,000 well-equipped hackers in the United Kingdom, that they were "widespread and lethal" and that these "evil doers" could only be properly dealt with were computer hacking made into a specific criminal offence.[1] She had been a computer programmer before entering Parliament and, like the Earl of Northesk (see Chapter 8), she argued that one of the main reasons that this problem was not being properly dealt with was the technical ignorance of parliamentarians. She pointed out that Britain's European

DOI: 10.4324/9781003252504-10

neighbours and competitors, such as the French, already had legislation in place and that were the problem not tackled in Britain, then the country might lose business to this, more protected environment. Nicholson's paper was full of apocalyptic scare stories gleaned from popular media. She alleged that a 1986 hack of the systems at CERN (which she mistaken called a nuclear reactor) could have triggered a strike by the United States' "star wars" missile defence system, that the German Green Party had bought computers to hack the country's census and that the Japanese government was funding an international hacking operation.[2]

Such stories were quite commonplace in the period. For example, three years before Nicholson initiated her crusade, the data protection registrar commissioned a report to assess the veracity of stories, about what was then termed "electronic eavesdropping," that were doing the rounds in the media. These included reports broadcast in a TV programme presented by James Burke, a self-appointed scientific expert, in which he alleged that by using only widely available equipment, he had been able to detect the radiation emanating from a government building in the north east of England. This was assumed to be the pension's office, and the programme implied that this could be unscrambled in order to read the original text. Similar stories were printed in the magazine *Computing*, which, in 1986, claimed that its reporters had been shown text derived by picking up the radiation emitted by computers in a branch of a building society. This article also alleged that similar experiments had been successfully conducted outside branches of other commercial premises. The registrar's report gave all these stories short shrift and suggested that, even if such radiation could be picked up, the simple tactic of turning two monitors on at the same time would utterly confuse anyone picking up the emissions and hence solve the problem.[3] Nevertheless, this was the febrile atmosphere into which Nicholson stepped and she was able to mobilise support from among her own party for her proposals, while the Home Office, perhaps unsurprisingly, favoured her proposal of legislation to make hacking a criminal offence.[4]

This furore caused two sets of problems for the government. The first concerned its immediate response to Nicholson. This was made more complicated by the fact that the Law Commission was already at work considering precisely this matter. Lord Young, at the Department of Trade and Industry, agreed with Nicholson's presentation of the costs of the problem but urged Major to wait and see what the Law Commission concluded whilst suggesting that the home secretary might chivvy that body along to report more quickly.[5] This, he argued, would demonstrate to all concerned that the government was taking the issue seriously. In the end, the government agreed to allow Nicholson's private member's bill to run its course unblocked by the government in Parliament, to thank her for her work in highlighting the issue and to pledge to bring forward its own legislation at some unspecified point in the future. The second, and potentially more far-reaching, problem this produced was that it threw the spotlight, albeit tangentially, onto the government's own computer operations. It was pointed out to Major, that the government had two roles in this furore. The first was to provide an appropriate legal framework for computer security, but it was also duty bound to provide an appropriate role model for others to emulate. Government departments were major users of computers; thus, it was "important for civil service departments to review their computer security arrangements."[6] The problem was that whenever these arrangements were reviewed, they were usually found to be wanting.

In 1971, as the government prepared itself for the Younger Committee's report on privacy, it conducted an enquiry into the security of government paper-based records. This was undertaken because the government wanted to reassure people that computer-based systems, the subject of Younger's investigations, were every bit as safe as government manual-filing systems. However, what the investigation unearthed unsettled the officials involved. They looked at the records of a variety of departments covering tasks selected to span the full breadth of the confidentiality spectrum. These ranged from the registration of births, marriages and deaths and the registration of driving instructors at one extreme, to prisoners' records and details of cancer treatment at the other. This investigation found that the systems in place to prevent unauthorised access to these records were, in most cases, inadequate. The records of births, marriages and deaths, the least sensitive material they examined, were stored in underground rooms, whilst the records of the Prisons Department in the Home Office were, more or less, open to anyone with access to the department. They were thus obliged to conclude that: "at the risk of stating the obvious," they had found "lax internal security" procedures.[7] Moreover, the investigators only found one instance where these practices had been changed by the introduction of computerisation. This meant that the security of computerised data was treated in the same slipshod manner. As though this were not bad enough, it also found that computer staff were not specially trained in security and neither were they subject to any special vetting procedures.[8]

It might be thought that such a lamentable state of affairs was inevitable during the introduction of computerisation and that things would have improved as computers became more common across Whitehall and officials became commensurately more clued-up about the technology. But reports from later years show that this transformation in attitudes did not occur in tandem with a transformation in data-storage methods. Thus, eleven years later, a survey of official attitudes to computer security showed that staff were uncertain about how to report a security incident, were unaware of how to respond to a virus attack and that half of staff, at all grades, thought that getting the system up and running was more important than ensuring it was running securely. Later surveys would also show that staff were, in many senses, caught in a dilemma of both needing to run systems as quickly and efficiently as possible and the drag that security measures would place on achieving this. For this reason, it was widely understood that security worked best when it was designed into a computer system from its inception, rather than being retrofitted as it often was. But security cost money and though the results of a substantial loss of data could be catastrophic, the actual chance of it occurring seemed remote.[9] On this balance of probabilities, security measures would always be applied incrementally; they would, in other words, always be the poor relation. Furthermore, it is worth noting that this survey concluded that one immediate action that ought to follow its publication was that every person who used a computer should have a copy of the manual that covered general security and the correct use of passwords, which is to say that, at the point of surveying, they did not have this.[10]

Five years later, in 1987, a report by the National Audit Office (NAO) argued that computerised data needed to be more protected from system failures than from hacking. But it found that officials had been very slow in responding to requests from the Central Computer and Telecommunications Agency to ensure that data was properly backed up. In fact, one-quarter of sites had no plans to do so at all, while a further third had only partial schemes. Of the plans that did exist, only a quarter had been properly documented and only one-eighth tested properly. Eight out of thirty-seven sites surveyed were unsupported and six were only partly backed up.[11] The following year, 1988 (the year before Nicholson poked the hornet's nest), the House of Commons Public Accounts Committee reiterated these points recording its "surprise and concern" that more action had not already been taken by departments in response to the NAO report's exposure of their vulnerability to serious loss of data.[12] In 1991, the NAO revisited this issue and found that though most departments had appointed an IT security officer and had established committees to oversee IT security, the Department of Social Security (DSS), always a big user of both data and computers, had actually disbanded its security oversight committee.[13] But even where these higher-level bodies were in place, it remained the case that, at the grass-roots level, staff awareness of security issues was "generally poor," with the NAO reporting that in local DSS offices, security was both "inconsistent and inadequate."[14]

In the early 1970s, officials had sought to comfort those concerned about the security implications of computerisation. This was done by stressing that digitised data was protected by the facts that the technology was beyond most people's understanding and that the information within the system was stored on tapes that were kept under lock and key away from the machines that could read them.[15] By the early 1990s computers were, partly as a result of the government's own efforts (see Chapter 8), common enough to feature on prime-time TV programmes, so the reassurance offered by the arcane nature of the technology could no longer really hold much water. Moreover, the same conclusion applied to the idea that the machines themselves, or their datasets, were physically inaccessible behind locked doors. Thus, the 1991 NAO report found that access to key sites in the DSS was, "poorly controlled."[16] Nevertheless, as was seen in Chapter 7, when pushing his ID card initiative to John Major, Michael Howard had been keen to emphasise the safety of the system underpinning a government smart card, a view that was echoed by the government's response to comments it received on *government.direct*.[17]

2

These reports gave the lie to government's statements that in giving information to its departments, the public were placing it in safe hands. As has been seen (see Chapter 5, for example), there was not any difference between data held on paper and that held on computer, and for the purpose under discussion here, whether this information was stored, retrieved and manipulated manually or electronically did not make much difference, since security was equally bad for both. But looked at from another angle the fact that computers were used did make a difference, because government had been obliged to make some concessions to British political attitudes. These came in the forms of the Younger Committee and the 1984 Data Protection Act, both of which the government had structured to ensure they focused on computers precisely to allow it to cover its data-gathering actions with a veil of technology. Indeed, as was seen in Chapter 6, the government increasingly used the technological nature of its systems for linking data as the basis for its reassurances to the public. All of which, notwithstanding the public nature of the concessions it had made, allowed for the growth of the depoliticisation that was at the heart of the data turn in government. Moreover, computerisation mattered because it was clearly the direction of travel for government data systems. Indeed, as was seen in Chapter 8, government was determined that this would be the course followed by the country more generally and launched programmes to aid and accelerate this societal change. The data turn was a political project, conceived at a time when computers were so primitive they actually processed data more slowly than traditional paper-based, systems.[18] Nevertheless, it was a political project based on data, and as such, those driving it would seize any opportunity offered by the new data-processing technology once this became sufficiently advanced to do anything other than the most routine administrative tasks.

The computerisation of government population-data systems was thus an important element of the developments covered by this book. But since this was data about the British people, it is important to gauge the nature of peoples' attitudes to the use of this technology for harvesting and holding their personal information. In 1989, the data registrar engaged researchers to do just this, and the rest of this section is based on this research.[19]

This was a qualitative-research project conducted through a series of twelve focus groups, each of which was about three hours long and involved a total of ninety-two participants in three locations. The people involved were chosen to represent the age, socio-economic and gender profiles of the population and crucially, were also selected to represent the full range of concern about protecting their own personal information. The data registrar's powers were confined to computerised data in the private sector, but this research wanted to understand the public's attitudes in their broadest sense and so enquired into all forms of data collection by both the private and public sectors. In the initial phase of the project, participants were accompanied to a variety of locations to apply for a passport, a loan, housing benefit (a social-security payment), household insurance, credit to buy household goods or a railcard. This experience then became a springboard for discussions about how many people/organisations might subsequently be able to view the data they had provided during these everyday exercises. These discussions revealed that:

> no-one was aware of the full extent of and scale of secondary data usage [by which the researchers meant usage over and beyond the reason for which the data had been supplied in the first place], even in the private sector, and there was overall very little awareness of information management in the public sector ... understanding of information technology was simplistic.[20]

Researchers grouped the participants into those who had very low, medium or quite high levels of awareness about the secondary uses to which their data might be put: tellingly none of the participants were placed in the high-awareness group. People thus generally expected to be asked for personal details; however, they had a limited understanding of how far and wide this information would be spread. On this basis, researchers recorded reactions of shock from participants when they were told that credit reference agencies routinely checked people with a similar name or address to an applicant for credit and that these agencies kept data on huge numbers of people. Apparently, most participants seemed to think that agencies functioned by contacting a reliable person (an employer, for example) and simply asking whether the applicant was a financially sound and generally trustworthy person.[21]

This lack of understanding might be seen as a reason to position these people somewhere on a spectrum running from naivety to indifference, or perhaps as indicative of the historical novelty of the commercialisation of data in the late 1980s. However, government was not ignorant of these trends; in fact, it was the year before this research was conducted that the Pickford Report referred to population data as the "fourth resource of a modern economy."[22] Moreover, the government was not only encouraging these developments in the private sector, but it was also actively participating in such commercialisation itself. Thus, it was only four years after this research that the data registrar would enquire into National Health Service (NHS) contract data sets (see Chapter 5). That this later enquiry into the NHS was necessary was highlighted by the fact that though most participants understood their information would be exchanged between medics, "it was widely believed that such information would not be disclosed outside the NHS."[23] In other words, what might appear to be naivety or indifference was in fact a product of people not being told what was happening to the information they gave to commercial or government organisations. Moreover, it would also seem that in initiating the registrar's enquiry into NHS datasets, the Joint Consultants' Committee was representative of broader attitudes current in society at the time.

Another facet of population-data systems that the people were largely kept in ignorance of was that their data, once provided for any reason, was highly likely to be linked to other sets of information. Indeed, many of the participants in this project were largely unaware that this was even possible. Thus, researchers noted that the participants "had no idea of the ways in which databases could be cross-referenced" and did not know that machines could, for example, add a postcode to an address.[24] Given this lack of knowledge, most of the participants expressed very little initial concern about the ways in which their data might accrue to construct what Rayner had called "coral reefs" within data systems (see Chapter 4). Most of these people had assumed their details would have been recorded on between five and fifteen data systems. But by working on the various real-life examples they had been accompanied through at the onset of the research, and working out how these different applications could cross-refer to each other, it became clear to most that they were probably listed on at least fifty different databases. This was, "everyone admitted … far more than they had imagined."[25] Similarly most of these members of the public were initially in favour of ID cards. This was particularly true of those participants who were old enough to remember having an ID card during the Second World War. There was very little principled opposition to these. The only voices raised against them were opposed to having to carry and produce them on demand, not to actually having one. Most participants thought that an ID card would be like those that had been issued in wartime: a simple piece of card bearing the holder's name and address. Thus, when they were told that, were they introduced, ID cards might be smart cards, most people were perplexed and unable to comprehend what this might mean, or why anyone would even want to introduce such a thing.[26]

It was at this point in the focus group discussions, after participants had been enlightened about what data gatherers had been doing and the prevalence of their systems for compiling information that they were each asked to produce a sketch to express how these revelations made them feel. They were also asked to write an accompanying sentence or two to explain what they had thus presented. The researchers made it clear in their report that these were produced after participants had worked out the scale of data gathering, they did not turn up to these focus groups already feeling like this. On the one hand, this might be taken to mean that these drawings represented people who had been disorientated by the researcher's shock tactics. But on the other hand, it needs to be noted that the researchers did not expose these members of the public to much detail of what the government itself was up to with its data-linkage systems, probably because these schemes were all under wraps. In other words, if these participants were shocked by what they deduced, they could probably have been yet further shocked. Moreover, they were not presented with anything that was untrue, and though it was not presented in a positive light, neither was it tarnished. Indeed some participants accepted these developments as being wholly positive.

Thirty-one of these drawings were presented in an appendix to the main report. One of these evidenced the disinterest of the person who described a drawing thus: "I'm relaxing in my own back yard and I don't care. As long as it doesn't affect me I really don't care." While another one saw these developments as something to be pleased about: "that's me up in the trees. I'm as free as a bird, and all the criminals are in prison because they got caught out by their ID."[27] Others stated that they were not concerned because they were too young, or too old, to be impacted by these issues. However, a large majority "were concerned and expressed emotions and fears of being spied upon, de-humanised, categorised and traded."[28] The report of the project classified ten themes prevalent in these responses all of which capture the raw essence of what the Lindop Report (see Chapter 5) had described as a "considerable threat to the privacy, and perhaps the freedom of, the private citizen" which "would greatly reduce the British citizen's traditional anonymity."[29]

The first of these was a feeling of being watched or spied upon. As one participant wrote to accompany a drawing: "I've got me standing in the middle of a room full of eyes asking me questions." The second theme was one of small, frightened individuals caught in a Kafkaesque nightmare, while another common thread was a reference to "1984." Representative of the former was the person who wrote: "absolutely helpless, you can't get away from it. Because there's no law to back you up is there?" While for the Orwellian version, one wrote: "Big Brother, plastic, watching you, me little person, versus the power and enormity of the computer." Two other themes were those of people being prevented from achieving their ambitions, being trapped, or a fear of being excluded from aspects of day-to-day life by a decision that they had no hand in making and against which they had no appeal. Thus, one respondent wrote of a fear of: "never being able to own what I want," while another wrote of being asked to show an ID "before you were allowed to get into a shop in case you were the type of customer who didn't pay up." A sixth theme was of being ordered around: "in my picture it's the computer giving instructions ... there are no other humans, the computer gives the orders directly." This last drawing was explicitly about the new technology and so were the last four of these themes from the respondents' drawings and writings. These were a fear of being classified, numbered and stereotyped; a fear of being trapped by the information held on computers; concern about computers sending information to each other, ignoring the individual concerned and a fear of computers holding erroneous information that then became established as the truth. With this last topic one person wrote of their drawing: "I'm saying 'I don't believe this is happening to me, you must have made a mistake', and the computer's saying 'We don't make mistakes.' And I'm saying 'How can I prove I'm innocent?' When the computer's saying 'You can't.'" For the other points people drew grids or cages, or "the government computer because if it doesn't exist now, I'm sure it will soon. That's me in the corner [a tiny figure] going 'help!' I'm dominated by this network of machines."[30]

It is worth noting that this research was conducted in the spring of 1989, which is to say it was almost exactly a year after the Government Data Network had been presented to Parliament (see Chapter 8). In other words, the hardware for the system that these members of the public clearly regarded with such loathing and dread was, by the time they communicated these feelings, already in place. This was also a year after Pickford reported on the Government Statistical Service (GSS) and so accelerated the centralisation of the government's data-gathering institutions. Moreover, this was only seven years before the government started planning the *government. direct* system. When this system was discussed in the previous chapter, it was shown how the government sought to present it as a means for interacting with people as "real world objects" that would make their day-to-day interactions with the government, "more accessible, more convenient, easier to use, quicker in response and less costly."[31] In doing so, the government poured scorn on the idea that the system would in any way whatsoever conform to the fears of the participants in the 1989 research project, such fears were said to be "utterly groundless."[32]

But this dismissal could not make such fears vanish and the fact that they did not disappear meant that the government was left with the problem of how to navigate them. As was seen in Chapter 5, its chosen twin-track solution was to emphasise the consumerist-type benefits (an easier life) that would flow from its data operations, and to direct the public's attention at the technological aspects of these systems where government could offer explanations and reassurances safe in the knowledge that people would lack the insights and skills to question them. Following on the discussion in the previous chapter, it is now possible to see that government may have hoped that as people became more familiar with the technology, they would also become less fearful of it. However, as was seen in Chapter 4, the spread of IT also opened new methods by which fraudsters might access data and so, whilst perhaps dampening down one aspect of the public's fear, the spread of IT across society stoked up another. What this means is that the one thing that was missing from the government's political relationship with the people was serious dialogue about the data it held. In the previous chapter, it was noted that: "reading parliamentary acts is not a common pastime." In a similar vein, these researchers pointed out that it was expecting far too much of people to assume that they would read the "small print" about how their data would be held, used and shared every time they passed on their details during a commercial transaction.[33] But, the government did not even offer such small print. Under these circumstances, the report concluded that: "there was a strong demand for general publicity – from all types of data users."[34] The previous chapter showed that the government did indeed issue publicity concerning IT, but it was not designed to achieve this sort of serious engagement. Since this political engagement was not forthcoming, the government would, as Jürgen Habermas noted, continue to present these serious political issues as technical problems, capable of being fixed by technical means, while simultaneously being fully aware of the paucity of its own data security provisions.[35]

Chapter 5 showed how, as a result of popular agitation around the issues of personal privacy, the government appointed the Younger Committee and eventually, at the behest of the EC passed the 1984 Data Protection Act. This act only applied to data held on computers in the private sector and as such was part of the policy vector pursued by the government of defining its population-data gathering as a purely technical exercise. But shortly after Tony Blair became prime minister, the government published new proposals on data protection. How these proposals were engendered and framed, and thus how they fit into this broad pattern of government relations with the British people concerning population-data gathering is the subject of the next section.

3

In September 1990, the Commission of the EC issued a communication *On the Protection of Individuals in Relation to the Processing of Personal Data in the Community and Information Security.*[36] This document argued that there were three reasons why the EC and its members both needed and wanted to enhance their level of data protection. First, EC integration needed it. As states became more closely aligned across a wider range of socio-economic areas, they inevitably needed to exchange more data, and this could be aided by the harmonisation of existing data protection measures. Indeed, it argued that a failure to do so could easily become a barrier to both free trade and the free movement of people within the community. Second, there was growing pressure from within Europe, from the European Parliament, for example, for improvements in the protection offered to ordinary citizens. Third, there was the new, and increasing, challenge posed by automatic data-processing and communications networks, both of which made it easier to link data sets and thus called for enhanced protection against any technical failure or criminal activity.[37] The Commission argued that such protection ought to cover state activities, not only those of private-sector organisations, and that when it were drafted, it might well seek to cover all datasets, including paper-based files. The European legislation thus sought to achieve two things simultaneously: to extend data linkage and to bolster the ways in which this linked data was protected. This latter element of the package would cause problems in Britain. The House of Commons Home Affairs Committee welcomed the efforts of the European Commission to boost trade through harmonising community-wide data-protection legislation. However, when he gave evidence to this committee Eric Howe, the data protection registrar, fought a rear guard action in defence of Britishness and "expressed some concern over various practical difficulties which may arise from greater integration."[38] In other words, he thought that the British government would resist the Commission's proposals to radically overhaul and extend the level of data protection afforded to the British and instead would work to swing the balance further in the other direction: in favour of enhancing population-data stocks.

Howe was correct and, soon after their publication, opposition to these European proposals emerged within government and crystallised around two issues they raised. The first of these was the threat they seemed to pose to the gathering of population statistics. In March 1993, a meeting of the GSS's Committee on Social Statistics, chaired by Bill McLennan (he had become chief statistical officer the year before), heard how: "if the directive were made law then the collection of a large proportion of government statistics would be impossible," and recommended some strenuous lobbying in the corridors of Brussels to forestall this.[39] This seems to have borne fruit and a meeting of representatives from the statistical offices of all EC members, along with Eurostat and a selection of British government departments was held in London in July 1993. The British representatives at this meeting argued that British government data sets were particularly threatened by these proposals since the GSS was not a unified institution, but a "federal one, and so relied on data transfers being made across departmental boundaries."[40] More broadly, this meeting, unsurprisingly, produced a broad consensus against any restrictions that the Commission's proposals might place on the gathering of statistical data. It also produced discussions of the advantages to be gained by using administrative data for statistical purposes, linking data sets and employing common-numbering systems, all of which was very much business as usual for all those in attendance.

With this level of agreement on a topic so close to the hearts of all modern governments, and the EC itself, these statisticians would not find it too hard to get the exemption they sought for statistical data from the Commission's proposals. However, at the meeting, one of the British representatives raised a separate issue. This concerned the Commission's intention to include paper-based records within the scope of its rules. The meeting was told that "it would be a formidable problem" for the British government "to allow access to all manual data (e.g. 500,000,000 public health records alone)."[41] However, resisting this would pose the government far greater political difficulties than it had encountered in mobilising against the statistical element of the proposals.

British governments had worked hard to keep their paper-based records out of the public eye even to the extent of allowing their computers to continue to attract criticism and hostility in order to do so. But its antipathy towards opening its files of paper records in general, and in particular being told to do so by European institutions, had been brought to a head a little over a year before these proposals emerged. This came as a result of the European Court of Human Rights (ECHR) decision in the Gaskin Case.

Graham Gaskin was a British citizen born in Liverpool in 1959. Before he was a year old, his mother died and he was taken into care by the local authority. He spent most of the rest of his childhood, apart from a few brief stints with his father, living with a variety of foster carers in the city, until he fell foul of the law and was taken into custody. In 1979, he sought damages from the council for the abuse he suffered while under the care of foster families and, as part of this action, wanted to see the authority's records of his case. The council refused to grant this access on the grounds that people who had contributed to these records had done so after being told that the files would be confidential. The case went to the British High Court in February 1980 where Lord Denning ruled in favour of the council. Gaskin subsequently appealed against this decision but this was also defeated. Nevertheless, in 1982, Liverpool Council decided to open some of its files, subject to some redacting of medical and police information. One member of the committee overseeing this process dissented and the issue went to the DSS in London. From here, the government more or less reimposed the status quo of the Denning verdict on all files compiled before that date, but allowed for a Liverpool-type of disclosure for records compiled from that point on. Of course, this was no help at all to Gaskin since his records all predated this change of mind. It was against this decision that he appealed to the ECHR, and he was successful. The court found that the files held by Liverpool Council "related to the applicant's basic identity, and indeed provided the only coherent record of his early childhood and formative years."[42] Thus, denying him access to this information was to deny him an understanding of his identity and therefore deprived him of his right to a private life.

This verdict marked a watershed moment for the British system and its impact is still felt in laws governing access to personal social work records. However, what is more important for the discussion here is that, within a year of this decision, the European Commission produced proposals that must have seemed to the British government to be seeking to universalise the Gaskin decision and throw open all the government's paper-based files.

Less than a year after the proposals were first mooted, the DSS noted that its offices would face huge difficulties were claimants to be entitled to see all their records, many of which were still held on paper, and that the department "would prefer to see manual records excluded from the scope of the draft directive."[43] This became, and remained, the government's position. Thus, in July 1993, the United Kingdom's delegation to the working party on these proposals was instructed that: "the removal of all manual data from the directive is the priority."[44] It was this, plus the unacceptable costs that the proposals would place on business, that led the foreign secretary, Douglas Hurd, to write to the home secretary, Michael Howard, stating that Britain's "overall objective" was "scuppering the directive."[45]

The British government initially hoped to be able to muster enough support from other member states to engineer this scuppering. To do so, the government would need to mobilise a blocking minority and, in July 1993, it thought that by building a bloc with the Belgians, French, Germans, Greeks and Dutch, it might be able to do so. However, by December that year, this group's cohesion had collapsed and, by the following summer, Britain could only count on the support of Ireland and Denmark.[46] Thus, Britain's representatives were left diplomatically isolated and searching for a new strategy. Here the government had a choice, it could either "seek to obstruct negotiations so that no common position is reached before enlargement (in the hope of a more favourable balance of votes thereafter)," or it could compromise.[47] It would eventually opt for the latter course, but in the meantime, pursued the former.

The chief delaying strategy the government deployed to obstruct progress towards increased data protection was to exacerbate arguments about the precise definition of terms. The Commission proposed including paper records, or files, within the ambit of its rules and the British prolonged arguments over the definition of a file. Similarly, the Commission proposed that there be a transitional period applicable to paper files that were in use and the British sought clarification over precisely what was meant by the word use. Both these issues first appeared as bones of contention in December 1993 and were still being debated the following August when the British requested the matter be referred to legal council for some kind of clarification.[48]

The British government sought to justify its reluctance to embrace enhanced data protection in two main ways. First, it argued that the European proposals would have made the lives of ordinary people harder and more irksome. This was because the proposals initially sought to insist that data subjects should be asked every time any organisation holding their data wanted to either: use it for a purpose other than the one for which it had been collected, or, wanted to share it with, link it to or sell it to another data user (eventually the 1998 Act would introduce the familiar opt-in/opt-out clauses presented by a lot of data gathering). However, the reality was rather different from that presented by this apparent concern for taking up the public's time, and this was made plain by the negotiating brief given to members of the British delegation. "Our view," they were told, "is that there are cases in which proper consent for [data] processing is given implicitly by a general request for a service." This had been the reasoning underpinning the concepts of the "task" and the "episode" brought forward in the 1970s, in weakening traditional codes of confidentiality around health data, and by those seeking to justify *government.direct* respectively. Which is to say that the position in Britain, which the British delegation was told was "common sense," was that people were not told about how their data might be used, never mind asked whether they agreed to this.[49] Moreover, DSS officials claimed that, insofar as the proposals attempted to give people the ability to opt in/out of government's using their data for secondary purposes, they would be "disastrous" for the department. Additionally, were consent to be required for extra usage of data, this would, they argued, stymie research conducted on their stockpiles of data to the detriment of their activities, the government and the people.[50]

The second way the government justified its hostility to the proposals was to argue that they trampled on British traditions and sovereignty. Douglas Hurd and Michael Howard would disagree about the introduction of ID cards, but when Hurd wrote to his colleague arguing for scuppering the proposals, the two were on the same side. Howard was "convinced that we must fight hard to reduce the impact of the Directive. Too much is at stake for us not to do so."[51] He insisted that the proposal was "manifestly incompatible" with the EC's own policy of supporting subsidiarity and that this point of principle was, therefore, the best card the government had in its hand in these negotiations. As British officials told one of their German counterparts: "if the directive is necessary at all, it should be sufficiently flexible to allow member states to maintain and adopt national laws and procedures which best meet their particular circumstances."[52] It is worth pointing out here that Howard was, at this time, also in the early stages of agitating for the introduction of ID cards in Britain (see Chapter 7). As such, his stalwart defence of British political traditions only manifested itself when these traditions were threatened by European legislation designed to protect the British people from data harvesting. Such a devotion to Britishness was, of course, absent when it came to ID cards, and the same could be said for most of his colleagues, many of his predecessors and those who followed him in office. The view of Britishness they sought to defend was one where departments had carte blanche to use data as they saw fit, and this view, as this book has shown, was a tradition that had only quite recently been invented.

The Commission's proposals were eventually adopted, with the formalities being completed by the Council of Ministers in October 1994. The rules were to apply to paper records, but only to records made from that date onwards. The ruling would not be retrospective. The government issued a consultation paper about all this in March 1996, but putting these proposals into British law fell to the Blair government elected in May 1997, with the new law coming into force in July 1998, as part of which the data registrar was retitled and became the information commissioner.

The events around the passing of the British data law reveal a lot about the government's ideas. But they also demonstrate how far the British government had travelled since Wilson's revolution steered the country into a data turn in the 1960s. Thus here, in 1993, government could defend its opposition to European legislation on the grounds that this would prevent it using its stockpile of data to conduct research about the population. Such a statement would have been unthinkable thirty years earlier at which point government data was segregated into data islands only loosely navigated by a decentralised GSS and surrounded by reefs of confidentiality codes made more complex by the bars of institutional enclosures. The system Wilson inherited regarded government research as anathema, whereas the one bequeathed to Blair saw it as the lifeblood of a modern state. British attitudes to Europe clearly also changed as part and parcel of this overall process.

Thus if, by the 1980s and 1990s, Europe threatened to put a brake on British government population-data operations, in the earlier period, continental practices were seen as a beacon to follow and an exemplar to emulate. After 1964, British governments pursued policies that rapidly caught up with those on the European mainland as governments pursued the systems needed to standardise, centralise and link their data holdings. These systems could assume the overt form of ID cards, new registers for the poll tax, reforms of existing systems such as the electoral rolls, or they could come in the less obvious guise of new computer or common-numbering schemes. The variety of forms considered or used, and indeed the relentlessness with which these were discussed and deployed, speaks to the nature of the task at hand and the zeal with which government addressed this from the mid-1960s. Thus, by embracing a biopolitical worldview, the British government also set itself on a quest for a chimera: the pursuit of a perfect knowledge of the population. This data turn of British politics could serve a variety of policies: from Wilson's compassionate statistics, to the hostile environment directed against social-security scroungers and immigrants under later administrations. Because these policies were so different, it is easy to lose sight of the extent to which they were all data driven and how this information hunger loomed large inside governments from across the full width of the political spectrum.

Notes

1 TNA, PREM 19/2717, Emma Nicholson, *Paper on the Need for Computer Hacking Legislation*, 20 Mar. 1989, 1.
2 Ibid., 17.
3 TNA, JX 9/1, *A Review of the Current State of Electronic Eavesdropping and Ways of Minimising its Influence on Data Protection*, Dec. 1986.
4 TNA, PREM 19/2717, Gray to Walters, 16 May 1989, 1.
5 TNA, PREM 19/2717, Lord Young to Major, 6 Apr. 1989, 3.
6 TNA, PREM 19/2717, Gray to Walters, 16 May 1989, 3.
7 TNA, HO 411/17, *Government Computers and Privacy, a Report by CSD on Some Manual Systems of Personal Information*, Sept. 1971, 14.9.
8 TNA, HO 411/17, Government Computers and Privacy, *Meeting 12 Oct. 1971*, 4.
9 The National Audit Office, *The Management of IT Security in Government Departments*, 26 Feb. 1991, 1.

10 TNA, HO 524/5, *Security Awareness Survey Results*, July 1982, 2–3.
11 The NAO, *Computer Security in Government Departments*, 27 Oct. 1987, 9.
12 House of Commons, Committee of Public Accounts, *Computer Security in Government Departments*, 18 Apr. 1988, vii and viii.
13 The NAO, *The Management of IT Security*, 26 Feb. 1991, 15.
14 Ibid., 5 and 6.
15 TNA, HO 337/264, E4 Division Home Office, *Revised Draft Report to Ministers, Government Computers and Privacy*, n.d., ca. Mar./Apr. 1972, 10.
16 The NAO, *The Management of IT Security*, 26 Feb. 1991, 15.
17 TNA, PREM 19/4735, Howard to Major, 21 Sept. 1994, 2.
18 Claus. A. Moser, "The Future Role of the Central Statistical Office," *Statistical News* 38 (1977): 1.2.
19 TNA, JX 11/13, *Qualitative Research on Fair Obtaining for the Office of the Data Protection Registrar*, Mar./Apr. 1989.
20 Ibid., 13.
21 Ibid., 24.
22 TNA, CAB 164/1910/1, *Scrutiny of Government Economic Statistics*, 27 Sept. 1988, 10.
23 TNA, JX 11/13, *Qualitative Research on Fair Obtaining*, Mar./Apr. 1989, 56.
24 Ibid., 52.
25 Ibid., 26–27.
26 Ibid., 29.
27 TNA, JX 11/13, *Qualitative Research on Fair Obtaining for the Office of the Data Protection Registrar, 'A Picture is Worth a Thousand Words' An Appendix of Psychodrawings*, Mar. 1989, 8 and 17.
28 TNA, JX 11/13, *Qualitative Research on Fair Obtaining*, Mar./Apr. 1989, 14.
29 *Report of the Committee on Data Protection*, Cmnd. 7341 (Dec. 1978), 261 and 264.
30 TNA, JX 11/13, *Qualitative Research on Fair Obtaining*, Mar./Apr. 1989, 94–95; and, TNA, JX 11/13, *Qualitative Research, 'A Picture is Worth a Thousand Words'*, Mar. 1989.
31 *government.direct: Electronic Delivery of Government Services*, Cm. 3438, (Nov. 1996), Foreword, 1.
32 TNA, CAB 130/1516, Ministerial Group on IT, *Draft Statement: government.direct*, 25 Feb. 1997, 2; and, *government.direct*, Cm. 3438 (Nov. 1996), 16.
33 TNA, HO 411/17, *Government Computers and Privacy*, Sept. 1971, 1.1–1.7.
34 TNA, JX 11/13, *Qualitative Research on Fair Obtaining*, Mar./Apr. 1989, 14.
35 Jurgen Habermas, *Towards a Rational Society* (London: Polity, 1987), 103–104.
36 TNA, BN 109/639, Commission of the European Communities, *Communication 'On the Protection of Individuals in Relation to the Processing of Personal Data in the Community and Information Security'*, 13 Sept. 1990.
37 Ibid., 2–4.
38 House of Commons, Home Affairs Committee, *Annual Report of the Data Protection Registrar*, 12 Dec. 1990, vi.
39 TNA, RG 50/35, GSS, Committee of Social Statistics, *Meeting*, 5 Mar. 1993, 5.
40 TNA, BN 119/12/2, *EC Directive on Data Protection, Meeting of Representatives of European Statistical Offices*, 13 July 1993, 4.
41 House of Lords Select Committee on Science and Technology, *Agenda for Action in the UK*, 23 July 1996, 67; and, TNA, BN 119/12/2, *EC Directive on Data Protection, Meeting of Representatives of European Statistical Offices*, 13 July 1993, 4.
42 TNA, JX 10/1, *E.C.H.R., Gaskin Case, Judgment*, 7 July 1989, 13.
43 TNA, BN 109/639, Matthews to Hickson, 9 July 1991, 1.
44 TNA, BN 119/11/1, *Negotiating Brief*, 7 July 1993, 17. Underlining in original.
45 TNA, BN 140/508, Hurd to Howard, 18 Oct. 1993, 2.

46 TNA, BN 119/11/1, *Negotiating Brief*, 7 July 1993, 17; TNA, BN 140/508, *Draft EC Directive*, 7 Dec. 1993; TNA, BN 119/13/1, Harding to Carden, 29 July 1994, 2.
47 TNA, BN 119/13/1, Harding to Carden, 29 July 1994, 2.
48 TNA, BN 119/13/1, Morgan to Oldeman, 19 Aug. 1994, 1.
49 TNA, BN 119/11/1, Negotiating Brief, 7 July 1993, 14.
50 TNA, BN 119/11/1, Holmes to Sutton, 5 Mar. 1993, 1 and 2.
51 TNA BN 119/12/2, Howard to Hurd, 27 Aug. 1993, 2.
52 TNA, BN 140/508, *Draft letter to Dr Werner Teotmeier*, n.d., ca. Aug. 1993, 2.

Conclusion

Partly as a result of its own foot dragging and politicking in response to the European data protection proposals, the Conservative government ended up in a position where it could not even claim the credit for having passed the measure into UK law. This fell into the lap of the Labour government of Tony Blair that was elected in May 1997. This government had little choice about passing this law, and when it published its proposals, they were, to all intents and purposes, the same as those developed under John Major. Given the combination of Labour's more positive approach to Europe and the pressing need to pass something to satisfy European demands, this was all probably inevitable and so says little about the new government's own thinking. However, the fact that the Blair government chose not to proceed with its predecessor's *government.direct* programme might lead to the conclusion that this government had a different attitude towards population data. Though the details that would be provided by government papers for this period are not (at the time of writing) available, it is nevertheless the case that even a brief examination of some of this government's publications shows how such a conclusion is wrong. This chapter makes such a survey. It does this to demonstrate how this government put further impetus behind the six main developments that had been honed by all its predecessors since Wilson initiated Britain's modernising data turn. Hence it argues that Blair's was the latest in a line to pursue the same biopolitical agenda.

In Chapter 3, it was shown how, along with all the other governments studied here, Blair's found the population-data apparatus it inherited to be deficient and set out to revamp this bequest. This was made plain when, slightly less than two years after being elected, the government published a white paper described as "the hallmark of the government." This was called *Modernising Government* and evidences the first way in which Blair's project continued the data trends already in place when he took office.[1] *Government.direct* may have been abandoned, but its vision infused *Modernising Government*. Indeed, it could hardly be otherwise since though different parties produced these sets of proposals, they were products of the same biopolitical mindset and the data-driven view of policy that this engendered. "Modernising," in the sense that the tern was used in this hallmark

DOI: 10.4324/9781003252504-11

paper, thus meant bringing Britain ever more into alignment with practices that had, as Foucault remarked, long been common on mainland Europe. It meant establishing biopolitical interventions that would be both rooted in and driven by population data to protect or reform the people.

Thus, this paper was built around a phrase that would become one of the defining features of the Blair years: "joined-up government."[2] The developments proposed in *Modernising Government* were sought in order to introduce integrated policies that would address people's needs "in a joined-up way, regardless of the structure of government."[3] This gave the paper clear echoes of the whole-person approach of earlier years and similarly where earlier governments had used the terms "task" and "episode" here, in *Modernising Government*, the Blair government referred to "clusters" of "related government functions aligned to the needs of citizens."[4] Bringing the officials charged with delivering these clustered needs into alignment with each other was at the heart of the project. This, in turn, would rely on "consistent standards" being applied across government and ensuring "that all the government bodies with an interest in a particular set of services come together to talk to potential partners, and that they promote compatibility across IT systems and data sets." Departments were to be encouraged, "to present the data they already hold in a common ['consistent and standardised'] way."[5] In other words, in modernising the British system of government, Blair was determined to drive through a broadening and deepening of the links made across government data. Through doing this, the government would seek to create a system that offered what a 2003 white paper would describe as a "living record," rather than a series of "snapshots."[6]

This is exactly what modernising had meant to Acheson, Titmuss and Moser in the early years considered in this book and to Redfern and McLennan in the later period. As such, it shows how modernity should be seen the way Foucault defined it, as "a practice," the sum of the methods used to solve the problems raised once population became the focus of thinking for those in government.[7] In attempting to deepen the links between government data sets, Blair was following in his predecessors' footsteps, however, their course had been necessitated by their inability to establish a system that would build the government's data anew. The abandonment of *People and Numbers*, the inability of Redfern to re-cast the electoral register and the failed attempts to introduce ID cards all necessitated government entering into a series of shotgun marriages with pre-existing institutions and data systems in an attempt to redefine and re-purpose the ways these functioned. This included circumventing existing confidentiality codes and assailing the institutional enclosures and the political/cultural attitudes of Britishness that they were based on. This tactical necessity was also why all of the governments examined here, and Blair's does not seem to have been an exception, drove through further standardisation and centralisation of the data they held. This standardisation of data, in the service of increasing the links between datasets, necessitated the centralisation of both data and the institutions that held, analysed and deployed it. This centralisation was promoted under Blair and represents the second way his governments both reflected and reinforced the developments initiated by Wilson's revolution. Indeed, given the desire for joined-up government, centralisation was inevitable.

But, at this point, the late 1990s, a desire to modernise also implied the increased use of technology to achieve this long-standing desire to link and centralise government population data and so, despite having jettisoned the schemes of the Major years, the Labour government, emphasised how IT could deliver these benefits of modern, linked government. This is the third way in which the hallmark of the Blair project was, in reality, a continuation of trends it inherited. Thus, the paper trumpeted "information age government" and declared: "we will use new technology to meet the needs of citizens and business, and not trail behind technological developments."[8] In this vein, *Modernising Government* repeatedly stressed how IT could be used to deliver government services that the people needed, where when and how these recipients needed them. This paper was published in the early years of the Blair governments so it did not have many examples of joined-up policy to present, the Sure Start programme (sixty pilot projects had been launched) was used but otherwise the government fell back on the same claim that its predecessors had deployed: that its plans would prevent people being asked to provide the same information every time they contacted a government department. Thus, like all its predecessors, the Blair government was unable, or unwilling, to envisage a modernisation programme that would treat people as anything other than consumers of public services or the objects of its data-gathering gaze.

This dismissive attitude was all of a piece with the nature of government population-data gathering as a mindset. Government was well aware that the norms of British political culture were inimical to the centralisation of government data, all of which was encapsulated in the word dossiers. Moreover, government was aware that though it did not matter to people whether such dossiers were on paper or computers, they were particularly concerned about the possible adverse impact of computerisation on their privacy. This was not an abstract understanding in government circles. Wilson's *People and Numbers* plan had been derailed by a privacy campaign and the 1981 census had been altered as a result of the Harringay protests. The people had made it obvious that they did not simply exist to be counted but rather wanted their opinions to count for something. However, none of the governments examined in this book engaged the people in a serious political dialogue about the data-driven means used to deliver policy and this included Blair's. As a result, apart from a cursory mention of the governments' commitment to devolution, there was nothing in *Modernising Government* about the empowerment of the people. In fact everything that was featured here, from what was proposed, to the technological methods of delivering these aims, can be seen as increasing the depoliticisation that had been characteristic of the politics of the data turn since its inception. This marked the fourth way in which this Labour government adopted and accelerated the biopolitical trends that ran through all the governments examined here since 1964.

On this point, it is important to note that this policy drift to data marked a very considerable departure from the government systems Wilson found when he entered Downing Street, those that characterised the traditions of Britishness, with its hallmark distaste for documents and which, as Foucault remarked, singled Britain out from practices on mainland Europe.[9] Thus, whereas in the early 1960s government and officials alike had actively resisted social scientists' requests for data, kept them as far away from policy-making as possible and bolstered the defences of confidentiality around sets of data, in its seventy-one pages, *Modernising Government* gave over a mere seventeen lines to a discussion of the protection of individuals' data.[10] The vast bulk of the white paper presented ways to continue the drive to modernise what were seen as Britain's archaic practices. Indeed, things had moved so far since the 1960s that the paper did not even mention the watchword of the previous era, balance, at all. Instead it stated that the government would conform to the existing data protection legislation and issued guarantees of technological safety against mistaken identity, inadvertent disclosure and inappropriate transfer of data. However, since the hallmark of the government, presented by the paper, was precisely the transfer and linkage of data, this latter point should be seen for what it was: a dismissal of the concerns of people who might have wanted to protect the political culture and practices of Britishness.

Though *Modernising Government* captured the zeitgeist of the Labour government, it was not the first time data linkage was raised by a government publication issued under Blair. This occurred in July 1998 when the government issued its white paper *Fairer, Faster and Firmer – A Modern Approach to Immigration and Asylum*. Here a "modern approach" was defined, in terms that were identical to both those used a year later in *Modernising Government* and to those used by preceding governments: one powered by the integration of the relevant services and the data streams they were both built on and deployed. In introducing the paper Jack Straw, the home secretary, echoed the language common throughout the period covered by this book and noted that the system he had inherited was a "shambles" based on a "piecemeal approach" that failed to coordinate services across government.[11] What was needed was to move towards a system, driven by IT, where sections of the immigration and asylum processes would share data. This, Straw stated, would allow for both the delivery of the eponymous faster, fairer and firmer system of decisions and also "better targeting of individuals and organised criminal groups who seek to abuse the system."[12] This unremitting desire to protect the government's data gathering from fraud marks the fifth main feature of the approach pioneered by its predecessors that was picked up and pursued under this government.

It was here, in this desire to protect its data systems from fraud, that it is possible to see the roots of why the Blair government, in common with its Conservative predecessors, was so keen not only to build a centralised data architecture, but also to identify people. To this end, in 2002, with David Blunkett as home secretary, this government would begin the process of introducing ID cards using, as was mentioned in Chapter 7, the same proposals, indeed even much of the same language, as Michael Howard. Moreover, it is worth noting that while the police had, to some extent, attempted to stand aside from the ID-card debate under earlier governments, under Blair's, they stated plainly that "a society built round an individual's true identity and their ability to prove it would significantly reduce the opportunity for crime in a number of areas."[13] This may have been the case, but such a society would also have ridden roughshod over the traditions of Britishness. It was, after all, representatives of the police who had, in 1996, described ID cards as being characteristic of a "police state" (see Chapter 7).[14]

The sixth aspect of its use of population data that the Blair government both inherited from its predecessors and subsequently embraced, was the way it used these enhanced systems to target the immigration issue in British politics. This is evidenced throughout *Fairer, Faster, Firmer* but can also be seen in how the government drummed up a popular feeling against what it termed "health tourists" in 2004.[15] These proposals, couched in the language of fairness and modernisation, were presented as necessary to "close loopholes" that left the Department of Health unaware of how many of the five million people who had registered with a GP in 2002/03 were from overseas and so "not properly entitled to free NHS primary medical services."[16]

After Wilson initiated the data turn in Britain, governments of all stripes sought enhanced knowledge of the population. In seeking this, they were keen to prevent people from being impervious to the gaze of data systems. Within this desire for all-encompassing knowledge of the population, this notion, highlighted by Straw, of preventing abuse, or fraud within the system, had been central to the population-data policies and systems of most of the governments studied here and this did not stop either when Blair came into office, or when *Fairer, Faster and Firmer* was published. In March 1998, the government published a white paper on the social-security system titled: *A New Contract for Welfare: Safeguarding the Social Security System.*[17] This made clear that the government's aim was to develop a system that was "secure from start to finish."[18] This security was to be garnered by making claimants produce original documents before they could claim benefits but also, more importantly, by realigning the Department of Social Security's data systems. In 1999, these systems operated in discreet units that filed data according to the benefits given, not by the individual receiving them. This meant that information about any individual claimant was not shared and that therefore, the social-security system as a whole was "a breeding ground for error and fraud."[19] Overcoming this structural weakness meant: "making intelligent use of information ... using to the full our powers to compare separate sources of information on any one individual to detect inconsistencies."[20] In other words, it would involve the mass linking of data.

In 1951, Lord Goddard had railed against the use of ID cards in Britain in a conservative defence of British political culture and though much the same data could flow from behind-the-scenes data linkage, ID cards remained the public face of population registers and so also of de novo data schemes to modernise British state practices. As was seen in Chapter 8, the Conservative government under John Major attempted to introduce ID cards and Blair's would follow suit in July 2002, when a government paper on the subject would be published hot on the heels of another on the issue of identity fraud.[21] That there was nothing really new in this policy is to be expected given that it was produced from the same set of attitudes as its predecessor and, for the same reason, the roots of Blair's ID-card project were already on display in the *Fairer, Faster and Firmer* and *A New Contract for Welfare* papers.

Thus, the former of these papers made it clear that "an effective identification system is an essential element of an effective removals system," while the latter expressed incredulity that, despite all Peter Lilley's fulminations, the system Labour picked up from the Conservatives did not have a "legal requirement for individuals claiming Income Support [a social-security benefit] to prove their identity as part of their claim."[22] Identification of the individuals that comprise the population is part of the gathering of population data so knowing the individual is an integral part of biopolitics. The problem inherent in this quest for knowledge, or identification, is similar to the enigmatic nature of modernisation: since it is a process, it has no predetermined end. Consequently, the knowledge already held can always either be parsed or reconfigured in new ways or it can be amended or supplanted by new knowledge. Thus, Jack Straw (above) stands as an example of how all the governments studied here lambasted the data they inherited from their predecessors in office and, in seeking to increase their panoptic view of the society on which they would act, sought to revamp data systems and gather more knowledge simply because there could always be more: they were engaged in the hunt for a chimera.

This book has attempted to chart the course of this data-driven turn in British politics and to show how the ideas that underpinned this development came to shape thinking across the party-political divide. However, this is not to say that this flow of ideas ran unimpeded along Whitehall. This book has considered the extent to which the government's data gatherers had to struggle to get what they wanted, through shotgun marriages with the structures that they inherited, and the ways in which they were diverted into dead ends and detours along the way. But, looking back from the vantage point of the 2009 Joseph Rowntree Trust Report at the way these events unfolded after 1964, it is clear that the British government travelled a very long way and our understanding of British political history is the poorer for not realising this. Moreover, given the nature of these developments, and the chimerical nature of what it is that the government's data gatherers seek, it is perhaps more important to understand that, from their perspective, Britain has not yet reached a destination and that the British data state will remain a work in progress.

Notes

1 *Modernising Government*, Cm. 4310 (Apr. 1999), 9.
2 Ibid., 5.
3 Ibid., 10.
4 Ibid., 53.
5 Ibid., 53 and 50.
6 *Civil Registration: Vital Change, Births, Marriages and Death Registration in the Twenty-First Century*, Cm.5355 (Jan. 2002), 25.
7 Michel Foucault, *Security, Territory, Population* (New York: Picador, 2004), 277.
8 *Modernising Government*, 13.
9 TNA, HO 310/314, Angel to Hart & Fittall, 25 Feb. 1987, 1.
10 *Modernising Government*, 51.
11 *Fairer, Faster and Firmer – A Modern Approach to Immigration and Asylum*, Cm. 4018 (July 1998), 3 and 20.
12 Ibid., 27.
13 House of Commons Select Committee on Home Affairs, *Fourth Report*, 30 July 2004 29.
14 House of Commons Home Affairs Committee, *Fourth Report, Identity Cards*, Vol.1, 26 June 1996, Minutes of Evidence, 23.
15 Department of Health, *Proposals to Exclude Overseas Visitors from Eligibility to Free NHS Primary Medical Services*, May 2004.
16 Ibid., 3.
17 *A New Contract for Welfare: Safeguarding the Social Security System*, Cm. 4276 (Mar. 1999).
18 Ibid., 9.
19 Ibid., 8.
20 Ibid., 3.
21 *Entitlement Cards and Identity Fraud: A Consultation Paper*, Cm. 5557 (July 2002), and Cabinet Office, *Identity Fraud: A Study*, July 2002.
22 *Fairer, Faster and Firmer*, 44, and *A New Contract for Welfare*, 7.

Bibliography

About, Ilsen, James Brown and Gayle Lonergan, eds. *Identification and Registration Practices in Transnational Perspective: People, Papers and Practices*. Basingstoke: Palgrave, 2013.

Acheson, E. Donald and John Grimley Evans. "The Oxford Record Linkage Study: A Review of the Method with some Preliminary Results". *Proceedings of the Royal Society of Medicine* 57 (1964): 11–16.

Acheson, E. Donald *Medical Record Linkage*. London: OUP and the Nuffield Provincial Hospitals Trust, 1967.

Adelstein, Abraham Manie. "Policies of the Office for Population Censuses and Surveys: Philosophy and Constraints." *British Journal of Preventive Social Medicine* 30 (1976): 1–10.

Agar, Jon. *The Government Machine: A Revolutionary History of the Computer*. London: Massachusetts Institute of Technology Press, 2003.

Alonso, William and Paul Starr, eds. *The Politics of Numbers*. New York: Russell Sage Foundation, 1987.

Anderson, Ross, Ian Brown, Terri Dowty, Philip Inglesant, William Heath and Angela Sasse. *Database State: A Report Commissioned by the Joseph Rowntree Reform Trust Ltd*. York: The Joseph Rowntree Reform Trust Ltd, 2009.

Beniger, James, R. *The Control Revolution: Technological and Economic Origins of the Information Society*. London: Harvard University Press, 1986.

Breckenridge, Keith and Simon Szreter, eds. *Registration and Recognition: Documenting the Person in World History*. Oxford: Oxford University Press, 2012.

Bulmer, Martin, ed. *Censuses, Surveys and Privacy*. Basingstoke: MacMillan, 1979.

Bulmer, Martin. "A Controversial Census Topic: Race and Ethnicity in the British Census." *Journal of Official Statistics* 2 (1986): 471–480.

Bulmer, Martin, ed. *Social Science Research and Government: Comparative Essays on Britain and the United States*. Cambridge: CUP, 1987.

Campbell, Duncan and Steve Connor. "The Battle against Privacy." *New Statesman*, May 9, 1986, 14–15.

Caplan, Jane and John Torpey, eds. *Documenting Individual Identity: The Development of State Practices in the Modern World*. Princeton: Princeton University Press, 2001.

Challis, Linda, Susan Fuller, Melanie Henwood, Rudolf Klein, William Plowden, Adiran Webb, Peter Whittingham and Gerald Wistow. *Joint Approaches to Social Policy: Rationality and Practice*. Cambridge: CUP, 1988.

Choldin, Harvey M. "Government Statistics: The Conflict between Research and Privacy." *Demography* 25 (1988): 145–154.

Crook, Tom and Glen O'Hara, eds. *Statistics and the Public Sphere: Numbers and the People in Modern Britain, c. 1800–2000*. London: Routledge, 2011.

Crosland, Charles Anthony Raven. *The Future of Socialism*. London: Jonathan Cape, 1964.

Curtis, Bruce. *The Politics of Population: State Formation, Statistics, and the Census of Canada, 1840–1875*. London: University of Toronto Press, 2001.

Dandeker, Christopher. *Surveillance, Power and Modernity: Bureaucracy and Discipline from 1700 to the Present Day*. London: Polity, 1990.

Desrosières, Alain. *The Politics of Large Numbers: A History of Statistical Reasoning*. London: Harvard University Press, 1998.

Donnison, David. "The Age of Innocence is Past: Some Ideas about Urban Research and Planning," *Urban Studies* 12 (1975): 263–272.

Dworkin, Gerald. "Reports of the Committees: The Younger Report on Privacy." *Modern Law Review* 36 (1973): 399–406.

Edgerton, David. "The 'White Heat' Revisited: The British Government and Technology in the 1960s." *Twentieth Century British History* 7 (1996): 53–82.

Egan, Mark. "Harry Willcock the Forgotten Champion of Liberalism." *Journal of Liberal Democrat History* 17 (1997–98): 16–18.

Flinders, Matthew and Jim Buller, "Depoliticisation: Principles, Tactics and Tools," *British Politics* 1 (2006): 293–318.

Foucault, Michel. *Discipline and Punish: The Birth of the Prison*. London: Penguin, 1991.

Foucault, Michel. *The Birth of the Clinic*. Oxford: Routledge, 2003.

Foucault, Michel. *Security, Territory, Population*. New York: Picador, 2004.

Foucault, Michel. *Society Must be Defended*. London: Penguin, 2004.

Foucault, Michel. *Psychiatric Power*. New York: Picador, 2006.

Foucault, Michel. *The Birth of Biopolitics*. New York: Palgrave MacMillan, 2010.

Foucault, Michel. *The Punitive Society*. New York: Picador, 2013.

Foucault, Michel. *Abnormal*. London: Verso, 2016.

Gamble, Andrew. *The Free Economy and the Strong State: The Politics of Thatcherism*. Basingstoke: MacMillan, 1988.

Goold, Benjamin, J. and Daniel Neyland, eds. *New Directions in Surveillance and Privacy*. Uffculme: Willan, 2009.

Habermas, Jürgen. *Towards a Rational Society*. London: Polity, 1987.

Hall, Peter. "Computer Privacy." *New Society*, July 31, 1969, 163–164.

Hibbert, Jack "Public Confidence in the Integrity and Validity of Official Statistics." *Journal of the Royal Statistical Society* 153 (1990): 123–150.

Higgs, Edward. *Life, Death and Statistics: Civil Registration, Censuses and the Work of the General Record Office, 1836–1952*. Hatfield: Local Population Studies, 2004.

Higgs, Edward. *The Information State in England: The Central Collection of Information on Citizens since 1500*. Basingstoke: Palgrave MacMillan, 2004.

Hoinville, Gerald and Terance Michael Frederick Smith. "The Rayner Review of Government Statistical Services." *Journal of the Royal Statistical Society, Series A* 145 (1982): 195–207.

Jeffreys-Jones, Rhodri. *We Know All About You: The Story of Surveillance in Britain and America*. Oxford: Oxford University Press, 2017.

Manton, Kevin. *Population Registers and Privacy in Britain, 1936–1984*. London: Palgrave MacMillan, 2019.

Margetts, Helen. *Information Technology in Government: Britain and America*. London: Routledge, 1999.

McLennan, Bill. "You can Count on Us – With Confidence." *Journal of the Royal Statistical Society. Series A* 158 (1995): 467–489.

Miller, Peter and Nikolas Rose, *Governing the Present*. Cambridge: Polity, 2008.

Moore, Peter Gerald "Security of the Census Population." *Journal of the Royal Statistical Society* 136 (1973): 583–596.

Moran, Michael. *The British Regulatory State: High Modernism and Hyper-Innovation*. Oxford: OUP, 2003.

Moser, Claus A. "Staffing in the Government Statistical Service." *Journal of the Royal Statistical Society* 136 (1973): 75–88.

Moser, Claus A. and I.B. Beesley. "United Kingdom Official Statistics and the European Communities." *Journal of the Royal Statistical Society* 136 (1973): 539–582.

Moser, Claus. "The Role of the Central Statistical Office in Assisting Public Policy Makers." *The American Statistician* 30 (1976): 59–67.

Moser, Claus A. "The Future Role of the Central Statistical Office." *Statistical News* 38 (1977): 1–6.

Oakley, Ann. "Fifty years of J.N. Morris's Uses of Epidemiology." *International Journal of Epidemiology* 36 (2007): 1184–1185.

OPCS Population Statistics Division. *Report of the Steering Committee to the Registrars General, Extending the Electoral Register – 1*. London: OPCS Occasional Paper, 20, 1981.

OPCS Population Statistics Division. *Extending the Electoral Register – 2, Two Surveys of Public Acceptability*. London: OPCS Occasional Paper, 21, 1981.

Pitt, Douglas C. and Brian C. Smith, eds. *The Computer Revolution in Public Administration: The Impact of Information Technology on Government*. Brighton: Wheatsheaf, 1984.

Pounder, Chris. "Police Computers and the Metropolitan Police." *Information Age* 18 (1986): 1–17.

Power, Michael. *The Audit Society: Rituals of Verification*. Oxford: Oxford University Press, 1997.

Redfern, Philip. "Population Registers: Some Administrative and Statistical Pros and Cons." *Journal of the Royal Statistical Society. Series A (Statistics in Society)* 152 (1989): 1–41.

Redfern, Philip. *Sources of Population Statistics: An International Perspective, Population Projections: Trends, Methods and Uses*. London: OPCS Occasional Paper, 38, 1990.

Redfern, Philip. "A Population Register or Identity Cards for 1992?" *Public Administration* 68 (1990): 505–515.

Redfern, Philip. "Precise Identification Through a Multi-Purpose Personal Number Protects Privacy." *International Journal of Law and Information Technology* 1 (1994): 305–323.

Rose, Nikolas and Peter Miller. "Political Power beyond the State: Problematics of Government." *The British Journal of Sociology* 43 (1992): 173–203.

Rose, Nikolas. "Governing by Numbers: Figuring out Democracy." *Accounting, Organizations and Society* 16 (1991): 673–692.

Rose, Nikolas. *Governing the Soul: The Shaping of the Private Self.* London: Free Association Books, 1999.

Rose, Nikolas. *Powers of Freedom: Reframing Political Thought.* Cambridge: Cambridge University Press, 1999.

Rule, James, B. *Private Lives and Public Surveillance.* London: Allen Lane, 1973.

Schofield, Camilla. *Enoch Powell and the Making of Postcolonial Britain.* Cambridge: CUP, 2013.

Scott, James C. *Seeing Like a State: How Certain Schemes to Improve the Human Condition have Failed.* London: Veritas, 2020.

The Royal Statistical Society. "Official Statistics: Counting with Confidence. The Report of a Working Party on Official Statistics." *The Journal of the Royal Statistical Society, Series A* 154 (1991): 23–44.

United Kingdom. Parliament. *Report of the Committee on the Provision for Social and Economic Research.* Cmd. 6868. 1946.

United Kingdom. Parliament. *Report of the Interdepartmental Committee on Social and Economic Research.* Cmd. 7537. 1948.

United Kingdom. Parliament. *Report of the Interdepartmental Committee on Social and Economic Research.* Cmd. 8091. 1950.

United Kingdom. Parliament. *Report of the Committee on Social Studies.* Cmnd. 2660. 1965.

United Kingdom. Parliament. *Report of the Committee on Local Authority and Allied Personal Services (The Seebohm Report).* Cmnd. 3703. 1968.

United Kingdom. Parliament. *Report of the Committee on Privacy.* Cmnd. 5102. 1972.

United Kingdom. Parliament. *Computers and Privacy.* Cmnd. 6353. 1975.

United Kingdom. Parliament. *Computers: Safeguards for Privacy.* Cmnd. 6354. 1975.

United Kingdom. Parliament. *Report of the Committee on Data Protection.* Cmnd. 7341. 1978.

United Kingdom. Parliament. *Government Statistical Services.* Cmnd. 8236. 1981.

United Kingdom. Parliament. *The Road User and the Law: The Government's Proposals for Reform of Road Traffic Law.* Cm. 576. 1989.

United Kingdom. Parliament. *Identity Cards: A Consultation Document.* Cm. 2879. 1995.

United Kingdom. Parliament. *government.direct: Electronic Delivery of Government Services.* Cm. 3438. 1996.

United Kingdom. Parliament. *The Government Reply to the Fourth Report from the Home Affairs Committee Session 1995–96*, Cm. 3362. 1996.

United Kingdom. Parliament. *Information Society: Agenda for Action in the UK. Government Response to the Report by the House of Lords Select Committee on Science and Technology.* Cm. 3450. 1996.

United Kingdom. Parliament. *Fairer, Faster and Firmer – A Modern Approach to Immigration and Asylum.* Cm. 4018. 1998.

United Kingdom. Parliament. *Modernising Government.* Cm. 4310. 1999.

United Kingdom. Parliament. *A New Contract for Welfare: Safeguarding the Social Security System.* Cm. 4276. 1999.

United Kingdom. Parliament. *Civil Registration: Vital Change, Births, Marriage and Death Registration in the Twenty-First Century.* Cm. 5355. 2002.

United Kingdom. The Cabinet Office, *Identity Fraud: A Study.* 2002.

United Kingdom. Parliament. *Entitlement Cards and Identity Fraud: A Consultation Paper.* Cm. 5557. 2002.

United Kingdom. Parliament. *Identity Cards: A Summary of Findings from the Consultation Exercise on Entitlement Cards and Identity Fraud*. Cm. 6019. 2003.
United Kingdom. Dept. of Health, *Proposals to Exclude Overseas Visitors from Eligibility to Free NHS Primary Medical Services*. 2004.
United Kingdom. The Cabinet Office, *Transformational Government Enabled by Technology*. Cm. 6683. 2005.
United Kingdom. Parliament. The Data Protection Registrar. Annual Report.
Wedgewood Benn, Anthony. *The Regeneration of Britain*. London: Victor Gollancz, 1965.
Wilson, Harold. *The New Britain: Labour's Plan Outlined by Harold Wilson*. London: Penguin, 1964.
Wilson, Harold. *Purpose in Politics: Selected Speeches by Rt Hon. Harold Wilson*. London: Weidenfeld and Nicolson, 1964.
Wilson, Harold. "Statistics and Decision-Making in Government – Bradshaw Revisited." *Journal of the Royal Statistical Society. Series A* 136 (1973): 1–20. Wilson, Harold. *The Governance of Britain*. London: Sphere, 1977.
Zuboff, Shoshana. *The Age of Surveillance Capitalism: The Fight for a Human Future at the new Frontier of Power*. London: Profile, 2019.
6, Perri; Charles Raab and Christine Bellamy. "Joined-Up Government and Privacy in the United Kingdom: Managing Tensions between Data Protection and Social Policy. Part 1." *Public Administration* 83 (2005): 111–133.
6, Perri; Charles Raab and Christine Bellamy. "Joined-Up Government and Privacy in the United Kingdom: Managing Tensions between Data Protection and Social Policy. Part 2." *Public Administration* 83 (2005): 393–415.

Index